Recent Trends in Plant Breeding and Genetic Improvement

Edited by Mohamed A. El-Esawi

Published in London, United Kingdom

Recent Trends in Plant Breeding and Genetic Improvement
http://dx.doi.org/10.5772/intechopen.111284
Edited by Mohamed A. El-Esawi

Contributors
Almir Karacic, Alp Ayan, Anandhi Krishnan, Ann-Christin Rönnberg Wästljung, Anneli Adler, Balaji Kannan, Basavaraj P.S., Bhojaraj Naik, Boraiah K.M., Budi Utomo, Burcu Gündüz, Deepa Bhadana, Halagundegowda G.R., Harisha C.B., Harsh Vardhan Singh Shekhawat, Hemant Sharma, Himanshu Pathak, Impa H. Ravindra, Ioana Stanciu, Jagadish Rane Sammi Reddy K., Jayusman , Jitendra Kumar Sharma, Kapil Choudhary, Krishnamurthy D., Lekha Lekha, Magnus Hertzberg, Martin Weih, Neeraj Kulshreshtha, Pratapsingh Khapte, Pär K. Ingvarsson, Rajaprakasam Sudhagar, Ramesh Kumar, Rami-Petteri Apuli, Shipra Singh Parmar, Sinan Meriç, Sivakumar Rathinavelu, Soheila Afkar, Sourabh Kumar, Subhash Chand, Sumaiya Sulthana, Tamer Gümüş, Tuğçe Aydın, Vaishnavi Vijayakumar, Vijayakumar H.P., Vijay Kamal Meena, Vijaysinha D. Kakade, Çimen Atak

First published in London, United Kingdom, 2024 by IntechOpen
IntechOpen is the global imprint of INTECHOPEN LIMITED, registered in England and Wales, registration number: 11086078, 5 Princes Gate Court, London, SW7 2QJ, United Kingdom

British Library Cataloguing-in-Publication Data
A catalogue record for this book is available from the British Library

Additional hard and PDF copies can be obtained from orders@intechopen.com

Recent Trends in Plant Breeding and Genetic Improvement
Edited by Mohamed A. El-Esawi
p. cm.
Print ISBN 978-1-83769-947-6
Online ISBN 978-1-83769-946-9
eBook (PDF) ISBN 978-1-83769-948-3

For EU product safety concerns:
IN TECH d.o.o., Prolaz Marije Krucifikse Kozulić 3, 51000 Rijeka, Croatia,
info@intechopen.com or visit our website at intechopen.com.

Meet the editor

Dr. Mohamed Ahmed El-Esawi is currently a visiting research fellow at the University of Cambridge, UK, and a professor at the Faculty of Science, Tanta University, Egypt. Dr. El-Esawi received his BSc and MSc from Tanta University and his Ph.D. from Technological University Dublin, Ireland. Afterward, Dr. El-Esawi joined the University of Warwick, UK; University of Sorbonne, France; and University of Leuven, Belgium as a visiting research fellow. His research interests focus on genetics, genomics, molecular biology, molecular physiology, developmental biology, and bioinformatics. He has been ranked among the world's top 2% of scientists. He has authored several international journal articles and book chapters, received several national and international awards and grants, and participated in more than sixty conferences and workshops worldwide. Dr. El-Esawi is currently involved in several research projects on biological sciences.

Contents

Preface

Plant breeding employs various approaches and technologies to enhance desirable traits in vegetables and crop plants. These approaches contribute to developing vegetables and crops with improved yield, resistance to diseases, and adaptability to changing environmental conditions, ensuring sustainable and resilient agriculture and food security. This book discusses the fundamental and current advances related to plant breeding and genetic improvement. It also highlights the physiological and molecular approaches employed to enhance plant tolerance to biotic and abiotic stresses. Moreover, it sheds new light on the future perspectives recommended to advance this field. This book is designed for plant breeders, researchers, scientists, and other interested readers, who find such information useful for enhancing their plant breeding technologies.

The book includes ten chapters. These chapters discuss the molecular selection tools in phenology of *Populus trichocarpa* breeds, stress memory and priming applications in plants, molecular marker technologies in plant improvement strategies, abiotic stress-tolerant crop varieties in India, cabbage culture and the possibilities of increasing the early production, accelerated approaches for cabbage improvement, allele mining and development of Kompetitive Allele Specific PCR (KASP) markers in plant breeding, breeding for macronutrient use efficiency (NTUE) in legumes, Indonesian toona breeding strategy, and the role of metabolites in abiotic and biotic stress tolerance in legumes.

The editor would like to thank the staff at IntechOpen, especially Publishing Process Manager Ms. Zrinka Tomicic and Senior Commissioning Editor Ms. Sandra Bolf, for their wholehearted cooperation in the publication of this book.

Mohamed Ahmed El-Esawi, Ph.D.
University of Cambridge,
Sainsbury Laboratory,
Cambridge, United Kingdom

Tanta University,
Faculty of Science,
Tanta, Egypt

Chapter 1

Advances in Molecular Marker Technology and Their Significance in Plant Improvement Strategies

Vijay Kamal Meena, Harsh Vardhan Singh Shekhawat, Subhash Chand, Kapil Choudhary, Jitendra Kumar Sharma and Lekha Lekha

Abstract

Molecular markers are powerful tools that have revolutionized plant improvement strategies by allowing breeders to select plants with desirable traits at an early stage. These markers are specific DNA sequences that can be used to identify genes responsible for important plant traits such as disease resistance, drought tolerance, and yield potential. Advances in molecular marker technology have greatly improved their efficiency and accuracy, making them an essential tool in plant breeding programs. One such advance is the development of high-throughput sequencing technologies, which allow for the rapid and cost-effective identification of large numbers of molecular markers. Additionally, new marker systems such as SNPs have been developed, which offer a high level of accuracy and reproducibility. The use of molecular markers in plant breeding has several advantages over traditional breeding methods. For instance, markers can be used to identify desirable traits that are not easily observable, or to select plants with multiple desirable traits at once. This has led to the development of new and improved crop varieties that are more resistant to diseases, better adapted to changing environmental conditions, and have higher yields. In conclusion, the continued development of molecular marker technology is crucial for the advancement of plant improvement strategies.

Keywords: molecular markers, markers assisted selection, crop improvement, genotyping, molecular plant breeding

1. Introduction

In the realm of modern agriculture and plant breeding, the integration of advanced biotechnological tools has revolutionized the traditional methods of crop improvement. Among these tools, molecular marker technology stands out as a pivotal innovation that has significantly accelerated the progress of plant breeding [1–3]. This technology allows breeders to gain insights into the genetic makeup of plants with unprecedented precision, facilitating the selection of desired traits in a more efficient and targeted manner.

Molecular markers are specific DNA sequences that can be easily identified and linked to particular traits or genes of interest. These markers serve as genetic landmarks on chromosomes, aiding in the identification of specific regions associated with advantageous characteristics such as disease resistance, yield potential, nutritional content, and environmental adaptability. By enabling researchers to pinpoint the genetic basis of these traits, molecular markers enhance the accuracy and speed of breeding processes [4–8].

Traditional plant breeding methods involve the crossbreeding of plants with desirable traits and the subsequent selection of offspring displaying those traits. However, this process is time-consuming and often requires multiple generations to achieve the desired results [6–8]. Molecular marker technology addresses these challenges by allowing breeders to identify and select plants possessing the desired traits at the molecular level, even before visible traits are expressed. This "marker-assisted selection" greatly expedites the breeding cycle, enabling the development of improved plant varieties within a shorter timeframe [9–11].

There are various types of molecular markers, including Restriction Fragment Length Polymorphisms (RFLPs), Simple Sequence Repeats (SSRs), Single Nucleotide Polymorphisms (SNPs), and Insertion/Deletion Polymorphisms (InDels), among others. Each type has its unique advantages and applications, making them valuable tools for different stages of plant breeding. With the advent of high-throughput sequencing technologies, genotyping a large number of markers has become increasingly feasible and cost-effective [12, 13].

Furthermore, molecular markers have not only expedited the breeding process but have also paved the way for more precise genetic manipulation through techniques like marker-assisted backcrossing and genome editing. This precision breeding approach ensures that only the desired genetic traits are introduced or modified, minimizing the unintended alterations associated with traditional breeding methods.

2. Classification of molecular markers

Molecular markers are crucial tools in modern plant breeding programs. They enable plant breeders to identify and select desirable traits with greater precision, efficiency, and speed compared to traditional breeding methods. Molecular markers are genetic variations that can be detected at the DNA or protein level. They are used to track specific genes or regions of the genome that are associated with desired traits. Here is a detailed classification of molecular markers used in plant breeding.

2.1 RFLPs (Restriction Fragment Length Polymorphisms)

RFLPs were among the first molecular markers developed. They are based on variations in DNA sequences recognized by specific restriction enzymes. Differences in the fragment lengths resulting from enzyme digestion can be visualized on an agarose gel. RFLPs are informative but require large amounts of DNA and are labor-intensive [14, 15].

2.2 AFLPs (Amplified Fragment Length Polymorphisms)

AFLPs involve selective amplification of DNA fragments using PCR (polymerase chain reaction). They are based on restriction enzyme digestion and subsequent ligation of adaptors. AFLPs generate numerous fragments that can be separated by gel electrophoresis, allowing for the identification of polymorphic markers [16, 17].

2.3 SSRs (Simple Sequence Repeats or Microsatellites)

SSRs are short DNA sequences consisting of tandem repeats (e.g., ATATATAT) that exhibit length polymorphisms due to variations in the number of repeats. They are highly variable, co-dominant, and have wide applications in plant breeding due to their abundance in the genome [18–20].

2.4 SNP (Single Nucleotide Polymorphism)

SNPs are single base pair changes in DNA sequences. They are the most abundant type of genetic variation and can be efficiently detected using various methods, such as microarrays or next-generation sequencing. SNPs are used in high-throughput genotyping and genome-wide association studies [21, 22].

2.5 CAPS (Cleaved Amplified Polymorphic Sequences) and dCAPS (Derived Cleaved Amplified Polymorphic Sequences)

CAPS and dCAPS markers are based on SNPs that create or abolish restriction enzyme recognition sites. After PCR amplification, digestion with a specific enzyme allows for the discrimination of different genotypes [23, 24].

2.6 RAPDs (Random Amplified Polymorphic DNA)

RAPDs involve random PCR amplification of DNA segments using short, arbitrary primers. These markers are quick and simple to generate but may lack reproducibility and require optimization [25, 26].

2.7 ISSRs (Inter Simple Sequence Repeats)

ISSRs involve PCR amplification between microsatellite sequences using anchored primers. They combine the advantages of SSRs and RAPDs and have been widely used for genetic diversity assessments [27, 28].

2.8 SNP arrays

SNP arrays are high-density microarrays containing thousands to millions of SNP markers. They provide efficient genotyping of large populations and enable genome-wide association studies and genomic selection [29, 30].

2.9 Genotyping-by-sequencing (GBS)

GBS is a next-generation sequencing-based approach that sequences subsets of a genome. It allows for simultaneous genotyping of many individuals at a reduced cost and is suitable for both well-characterized and non-model species [31, 32].

2.10 Methylation-based markers

These markers detect epigenetic modifications, such as DNA methylation. Methylation status can influence gene expression and phenotype, making these markers relevant for studying complex traits [33, 34].

2.11 Gene-specific markers

Markers can be designed to target specific genes of interest, aiding in tracking and selecting specific traits [35].

2.12 Transcriptome-based markers

Transcriptome sequencing can identify gene expression variations linked to traits. These markers are valuable for understanding gene function and regulation [36].

2.13 Exome capture markers

These markers focus on sequencing only the protein-coding regions (exons) of the genome, providing cost-effective genotyping information [37].

2.14 Copy number variation (CNV) markers

CNV markers detect variations in gene copy numbers among individuals, contributing to genetic diversity and phenotypic differences [38].

2.15 Insertion-deletion polymorphisms (InDels)

InDels are variations in the number of nucleotides within a specific genomic region. They are useful for distinguishing closely related individuals.

2.16 Sequence-tagged sites (STS)

STS markers are derived from specific DNA sequences associated with genes of interest. They can be used to directly amplify and analyze specific DNA fragments [39].

The choice of molecular marker depends on the specific breeding objectives, available resources, and the species under consideration. Advances in genomics and sequencing technologies continue to expand the range of molecular markers available to plant breeders, enabling them to accelerate the development of improved crop varieties.

3. Principles and techniques of molecular marker analysis

Molecular marker analysis has revolutionized plant breeding by providing tools to study and manipulate the genetic makeup of plants more effectively [40]. These markers serve as indicators of specific genes or genomic regions linked to desirable traits. Here's a detailed overview of the principles and techniques of molecular marker analysis in plant breeding.

3.1 Principles of molecular marker analysis

1. Genetic linkage: molecular markers are used to identify regions of the genome that are closely linked to target traits of interest. These markers are inherited along with the target genes, enabling breeders to select plants with desired traits through marker-assisted selection (MAS) [5, 41].

2. Quantitative trait loci (QTL) mapping: by analyzing the co-segregation of molecular markers and phenotypic traits in mapping populations, QTLs—genomic regions influencing quantitative traits—can be identified. This helps in understanding the genetic basis of complex traits [42, 43].

3. Genetic diversity assessment: molecular markers aid in evaluating the genetic diversity within a breeding population, ensuring that genetic resources are effectively utilized to develop improved varieties [20, 44].

4. Marker-assisted breeding (MAB): breeders can use molecular markers to expedite the traditional breeding process by identifying plants carrying desirable genes without having to wait for phenotypic expression [13, 45].

3.2 Techniques for molecular marker analysis

1. DNA extraction: isolation of high-quality DNA is the initial step. Various protocols are used depending on the plant species and tissue type [46].

2. PCR-based techniques: polymerase chain reaction is used to amplify specific DNA fragments using primers designed based on marker sequences.

3. Gel electrophoresis: amplified DNA fragments are separated by size using gel electrophoresis to visualize differences in marker patterns.

4. Hybridization: techniques like Southern blotting are used to identify RFLPs and other DNA fragments based on their hybridization patterns with labeled probes.

5. High-resolution melting (HRM): this technique identifies sequence variations by monitoring the melting behavior of PCR-amplified fragments [47].

6. DNA sequencing: sanger sequencing or NGS can provide the exact DNA sequence information for the marker region [48].

7. Genotyping platforms: SNP arrays and NGS platforms allow for high-throughput genotyping of thousands to millions of markers across the genome [49].

8. Data analysis: various software tools and statistical methods are used to analyze marker data, perform QTL mapping, assess genetic diversity, and make breeding decisions [50, 51].

9. Marker-assisted selection (MAS): breeders use marker information to select plants with desirable traits during the breeding process, improving efficiency and accuracy

10. Genome-wide association studies (GWAS): this technique involves analyzing marker data from diverse germplasm to identify associations between markers and phenotypic traits [52].

4. Applications of molecular marker technology in plant improvement

Molecular marker technology has revolutionized the field of plant improvement by enabling precise and efficient selection of desired traits in crops. These markers are specific DNA sequences associated with particular traits, and they provide a valuable tool for plant breeders to make informed decisions during the breeding process. Here, I'll cover various applications of molecular marker technology in plant improvement, along with examples of their use:

4.1 Marker-assisted selection (MAS)

This is one of the most common applications of molecular markers in plant improvement. MAS involves identifying markers linked to desirable traits and using them to select plants with those traits more accurately and efficiently than traditional methods [53]. For instance:

1. Disease resistance: breeding for disease-resistant plants by identifying markers linked to resistance genes. Example: development of rice varieties resistant to bacterial blight using *Xa21* gene marker [54].

2. Abiotic stress tolerance: identifying markers associated with drought, salinity, and cold tolerance to develop crops better suited for challenging environments [55, 56].

3. Genetic diversity and germplasm characterization: molecular markers help assess the genetic diversity of plant populations and germplasm collections, aiding in the conservation and utilization of genetic resources [57].

4. Fingerprinting: establishing genetic profiles of varieties to prevent misidentification and support intellectual property rights [58].

4.2 Phylogenetic studies

The use of molecular markers in phylogenetic studies has revolutionized the field by providing a more accurate and detailed understanding of evolutionary relationships among species [59, 60]. These molecular markers, such as RFLPs, SSRs, SNPs, and InDels, allow researchers to examine variations in DNA sequences [61, 62]. Identify unique genetic markers that can be used to determine relatedness and trace ancestral lineages. By analyzing these markers, researchers can create phylogenetic trees and reconstruct evolutionary histories. Furthermore, molecular markers have significantly improved our knowledge about past events that have shaped the genetic diversity within species [63].

4.3 Quantitative trait loci (QTL) mapping

The use of molecular markers in plant QTL mapping has revolutionized the field of plant breeding and genetics. By using molecular markers, researchers can identify and track specific regions of the genome that are associated with desirable traits in plants [61, 64, 65]. This helps to determine the criteria of selection and also combines molecular markers with conventional breeding methods to obtain the best results.

The progress in quantitative trait loci mapping in plants depends on the availability of DNA markers and efficient biometric tools, which lead to the accurate detection of QTLs [66, 67].

4.4 Marker-assisted backcrossing (MABC)

Marker-assisted backcrossing has emerged as a revolutionary approach in plant breeding, seamlessly integrating modern molecular techniques with traditional breeding methods. This strategy involves the use of molecular markers to identify and track specific desirable traits within the genome of a plant, facilitating the efficient transfer of these traits into elite breeding lines through a series of backcrosses. One of the most successful examples of this technique is the development of disease-resistant rice varieties. By identifying and incorporating markers linked to genes conferring resistance to devastating pathogens like bacterial blight and blast, breeders have rapidly produced high-yielding rice cultivars with enhanced disease resistance. Another notable achievement is the creation of drought-tolerant maize varieties. Through marker-assisted backcrossing, scientists have successfully transferred drought-responsive genes into commercially valuable maize lines, mitigating the impacts of water scarcity on crop productivity. These examples underscore the transformative potential of marker-assisted backcross breeding in addressing pressing challenges in agriculture and securing global food production.

4.5 Marker-assisted introgression of alien genes

Transferring beneficial genes from wild or related species into cultivated varieties to enhance traits like disease resistance, quality, or yield. Wheat rust resistance: incorporation of rust resistance genes from wild wheat relatives into cultivated wheat to combat fungal diseases [68, 69].

4.6 Genomic selection

Using genome-wide marker information to predict the breeding value of plants without having to observe the trait directly. This accelerates the breeding process by selecting individuals with the highest genetic potential. Animal forage quality: improving the quality of forage crops for livestock feed based on predicted genetic values [70–72].

4.7 Genetic mapping and genome assembly

Molecular markers aid in constructing genetic maps and assembling plant genomes, facilitating further research on gene function and evolution. Human consumption quality: enhancing the taste, texture, and nutritional content of fruits and vegetables for improved consumer satisfaction [73, 74].

4.8 Detection of genetic mutations

Identifying and characterizing mutations responsible for specific traits, diseases, or disorders. Seedless fruit development: understanding and manipulating the genetic mutations that lead to seedlessness in fruits like grapes and watermelons [75, 76].

4.9 Marker-based cloning of genes

Molecular markers help in identifying and isolating genes responsible for specific traits, enabling a deeper understanding of plant biology. Flower color modification: cloning genes associated with flower color to create ornamental plants with novel hues [77, 78].

4.10 Transgene detection and purity testing

Molecular markers can be used to detect the presence of transgenes in genetically modified (GM) plants and assess the purity of seed stocks. Bt cotton: identifying the presence of Bt toxin genes in cotton plants for insect resistance [79].

4.11 Marker development

Continual advances in molecular marker technologies, like Single Nucleotide Polymorphisms (SNPs) and high-throughput genotyping, allow the development of more efficient and cost-effective markers for specific traits [80].

In summary, molecular marker technology has profoundly impacted plant improvement by expediting the breeding process, increasing precision, and enhancing the ability to select and manipulate desired traits in crops. Its diverse applications have contributed to the development of improved crop varieties that address challenges ranging from disease resistance to environmental adaptation and nutritional content.

5. Marker Assisted Selection (MAS) in plant breeding

Marker-assisted selection (MAS) is a powerful technique in plant breeding that allows breeders to select plants with desired traits more efficiently and accurately by utilizing genetic markers associated with those traits. This approach has revolutionized the field of plant breeding by significantly speeding up the process of developing new crop varieties with improved characteristics. Here's a detailed explanation of marker-assisted selection, along with some examples of its applications.

5.1 What is marker-assisted selection (MAS)?

Marker-assisted selection involves the use of molecular markers, which are specific DNA sequences that can be easily detected and analyzed in a laboratory setting [81]. These markers are linked to particular traits of interest, such as disease resistance, yield potential, nutrient content, and other agronomic characteristics. By identifying the presence or absence of these markers, breeders can predict the phenotype of a plant and make informed decisions about whether to continue breeding or discard certain plants.

5.2 Steps in marker-assisted selection

1. Marker discovery: in this step, researchers identify and develop molecular markers that are associated with the desired traits. These markers can be Single Nucleotide Polymorphisms (SNPs), microsatellites (SSRs), or other types of genetic variations.

2. Marker validation: once potential markers are identified, they are validated across a diverse range of plant genotypes to ensure their consistency and reliability in predicting the desired trait.

3. Marker-assisted selection: after validation, the markers are used to screen and select plants during breeding. The presence of specific markers is used as an indicator of the presence of the desired trait.

4. Traditional breeding: the selected plants are then further evaluated through traditional breeding methods to confirm their performance and combine multiple traits.

5. Marker-assisted backcrossing: if needed, selected plants can undergo backcrossing, where they are crossed with a recurrent parent to recover their genetic background while retaining the desired trait.

5.3 Examples of marker-assisted selection

1. Disease resistance: one of the earliest successful applications of MAS was in developing disease-resistant crops. For instance, in rice breeding, markers associated with resistance to diseases like bacterial blight and blast have been used to identify and select plants with enhanced resistance, leading to improved crop yields [82, 83].

2. Quality traits: MAS has been employed to enhance the quality of crops, such as improving protein content in wheat and oil quality in canola. Genetic markers linked to these quality traits are used to identify and select plants with superior nutritional attributes [84].

3. Drought tolerance: with the increasing challenges of climate change, breeding for drought-tolerant crops is crucial. MAS allows breeders to identify plants with genetic markers associated with drought tolerance, enabling the development of varieties that can thrive in water-limited conditions [85, 86].

4. Yield enhancement: genetic markers associated with high yield potential have been utilized to accelerate the breeding of crops with improved productivity, helping to address global food security concerns.

5. Nutrient efficiency: breeders have used MAS to develop plants with improved nutrient uptake and utilization efficiency, resulting in crops with enhanced nutritional value and reduced fertilizer requirements [87].

6. Fruit traits: in fruit crops, MAS has been employed to select traits such as fruit size, color, texture, and taste, leading to the development of more appealing and marketable varieties [88].

7. Herbicide resistance: in crops like soybeans and corn, MAS has aided in the development of herbicide-resistant varieties by identifying genetic markers linked to resistance traits, allowing farmers to control weeds more effectively [89].

6. Genomic selection in plant improvement

Genomic selection (GS) is a cutting-edge plant breeding technique that leverages genetic information from an organism's entire genome to predict its potential performance and select the best individuals for breeding. This approach has revolutionized traditional plant breeding by allowing breeders to make more informed decisions and significantly accelerate the improvement of crop plants [90].

Genomic selection is based on the concept that the genetic information contained within an organism's genome can be used to predict its phenotypic traits, such as yield, disease resistance, and quality. This is achieved by analyzing a large number of genetic markers distributed across the genome, often in the form of Single Nucleotide Polymorphisms (SNPs). These markers are associated with specific traits, allowing breeders to create predictive models to estimate the potential performance of individual plants or animals.

6.1 Steps in genomic selection

1. Genotyping: the first step involves genotyping a large number of plants with molecular markers, typically SNPs, distributed throughout the genome. This creates a genomic profile for each individual.

2. Phenotyping: phenotypic data, such as yield, disease resistance, and other relevant traits, are collected from each plant in the breeding population.

3. Model training: statistical methods are then used to associate the genetic markers with the phenotypic data. This step involves building predictive models, such as genomic prediction models or genome-wide association studies (GWAS), which highlight the genetic markers associated with specific traits.

4. Model validation: the accuracy of the models is assessed using independent data sets. This helps ensure that the models are reliable and can predict the traits of interest accurately.

5. Selection: once validated, the predictive models are used to select plants or individuals with the desired traits for further breeding. This can greatly expedite the breeding process as only the most promising individuals are chosen for advancement.

6.2 Examples of genomic selection in plant improvement

1. Maize (corn): researchers have successfully applied genomic selection in maize breeding. By using a high-density SNP array, they created predictive models for

traits like yield, drought tolerance, and disease resistance. This has led to the development of maize varieties with improved performance under challenging environmental conditions [91].

2. Wheat: genomic selection has been used to improve wheat yield and quality. Researchers have identified genomic regions associated with traits like disease resistance, grain size, and nutritional content. By selecting wheat plants based on their genomic profiles, breeders have produced varieties with enhanced yield potential and nutritional value [92].

3. Rice: in rice breeding, genomic selection has enabled the development of varieties with increased resistance to pests and diseases. By identifying genetic markers linked to resistance traits, breeders can efficiently select plants that are less susceptible to yield-reducing factors [93].

4. Soybean: genomic selection has played a crucial role in soybean breeding. Researchers have employed this technique to enhance traits such as oil content, protein content, and resistance to pathogens. This has resulted in the creation of soybean varieties that meet the demands of both food and industrial applications [94].

5. Apple: in fruit tree breeding, such as apple breeding, genomic selection has been used to improve traits like fruit quality, disease resistance, and tree architecture. By selecting apple trees based on their genomic profiles, breeders can develop varieties that exhibit improved fruit characteristics and resilience [95].

These are just a few examples of how genomic selection has been applied to plant improvement across various crops. The technique's ability to rapidly and accurately predict the performance of plants based on their genetic makeup has significantly advanced the field of plant breeding, enabling the development of new varieties that are better adapted to changing environments, more resilient to diseases, and capable of meeting the needs of a growing global population.

7. Marker Assisted Backcross Breeding (MABB) in plant breeding

Marker-Assisted Backcrossing (MABC) is a powerful technique employed in plant breeding to accelerate the transfer of desirable traits from one plant variety (donor parent) to another (recipient parent) while retaining the genetic background of the recipient parent. This technique combines traditional backcrossing methods with modern molecular marker technology, enabling breeders to select plants carrying the desired traits more efficiently and accurately [96].

7.1 Steps in MABB

1. Trait identification: breeders identify a desirable trait, such as disease resistance, drought tolerance, or enhanced yield, that they wish to introduce into a particular plant variety.

2. Marker selection: molecular markers, specific DNA sequences associated with the desired trait, are chosen. These markers are polymorphic, meaning they

exhibit variations between different plants or varieties. Common types of markers include SSRs (Simple Sequence Repeats), SNPs (Single Nucleotide Polymorphisms), and AFLPs (Amplified Fragment Length Polymorphisms).

3. Donor parent selection: a donor parent possessing the desired trait is chosen. This parent serves as the source of the target trait and provides the genetic material needed to introduce the trait into the recipient parent.

4. Recipient parent selection: the recipient parent, usually a commercially valuable or locally adapted variety, is selected. This parent provides the genetic background that will be maintained throughout the breeding process.

5. Backcrossing: the donor parent is crossed with the recipient parent, resulting in the first generation of hybrids. The progeny from this cross is then backcrossed to the recipient parent for several generations. During each backcross, molecular markers linked to the target trait are used to select offspring that carry the desired trait.

6. Marker-assisted selection (MAS): molecular markers associated with the target trait are used to screen the progeny at each backcross generation. This allows breeders to select plants that carry the trait of interest while minimizing the incorporation of unwanted genetic material from the donor parent.

7. Selfing and selection: after each backcross generation, selected plants are self-pollinated to produce homozygous lines with a high proportion of the recipient parent's genome. These lines are then subjected to further rounds of selection using the molecular markers.

8. Genotype analysis: throughout the process, genotyping techniques are used to identify and confirm the presence of the target trait and monitor the proportion of recipient parent genetic material in the selected lines. Final Evaluation and Release: Once the desired trait has been successfully introgressed into the recipient parent's genetic background, the selected lines are subjected to extensive field trials and evaluations for various agronomic and quality traits. Once the lines meet the desired standards, they can be released as new improved varieties.

8. Marker Assisted Trait Mapping

Marker-Assisted Trait Mapping (MATM) is a powerful technique in plant breeding that involves the identification of genetic markers associated with specific traits of interest. These markers are then used to accelerate the process of selecting and breeding plants with desirable traits [97]. This technique has revolutionized plant breeding by enabling breeders to make informed decisions based on molecular data, leading to more efficient and targeted breeding programs.

8.1 Process of marker-assisted trait mapping

1. Trait selection: the first step in MATM is to select the trait of interest for improvement. This could be anything from disease resistance and yield to nutritional content and drought tolerance.

2. Marker identification: once the trait is chosen, researchers identify genetic markers that are linked to the trait. These markers can be Single Nucleotide Polymorphisms (SNPs), microsatellites, or other types of DNA variations. The idea is to find markers that are physically close to the gene responsible for the trait.

3. Population development: a diverse population of plants is developed for analysis. This population typically consists of individuals with varying degrees of the trait under investigation.

4. Genotyping: the DNA of each individual in the population is genotyped using techniques like Polymerase Chain Reaction (PCR) or high-throughput sequencing. This results in a dataset of genetic markers for each individual.

5. Phenotyping: the individuals in the population are evaluated for the trait of interest. This could involve measuring yield, disease resistance, or any other relevant phenotype.

6. Statistical analysis: the genetic marker data is statistically analyzed to identify associations between specific markers and the trait. This is usually done using methods like genome-wide association studies (GWAS) or linkage analysis.

7. Validation: the identified marker-trait associations are validated in different environments and populations to ensure their reliability.

8. Marker-assisted selection (MAS): once validated, these markers serve as tools for breeders. They can now use these markers to indirectly select plants with the desired trait in early generations, without waiting for the phenotype to become apparent. This significantly speeds up the breeding process.

9. High-throughput genotyping platforms in plant breeding

Plant breeding plays a crucial role in developing new crop varieties with improved traits, such as yield, disease resistance, and nutritional content. Genotyping, the process of identifying an individual's genetic makeup, is a vital component of modern plant breeding efforts. High-throughput genotyping platforms have revolutionized the field by allowing researchers to rapidly analyze thousands of genetic markers across plant genomes. These platforms enable the identification of genes responsible for desired traits and the development of markers for selection, accelerating the breeding process. This article provides an in-depth overview of high-throughput genotyping platforms used in plant breeding, highlighting key contributions from various researchers.

9.1 Polymerase chain reaction (PCR)-based techniques

Random Amplified Polymorphic DNA (RAPD): RAPD is a simple PCR-based technique that uses random primers to amplify DNA segments. Although less commonly used today, it paved the way for more sophisticated genotyping approaches. Amplified Fragment Length Polymorphism (AFLP): AFLP combines PCR with restriction enzyme digestion to generate DNA fragments for analysis. It was widely used in the late 1990s and early 2000s for plant breeding studies [98].

9.2 Microarray-based platforms

a. Genotyping by microarray: microarrays allow the simultaneous detection of thousands of DNA markers. Researchers design microarrays with probes that hybridize to specific target sequences. This technique has been used for genotyping crops like rice and maize [99, 100].

b. Diversity Arrays Technology (DArT): DArT involves probing genomic representations with a diverse set of DNA fragments. It is a cost-effective method for generating large amounts of genotyping data [101].

9.3 Next-generation sequencing (NGS)-based platforms

a. Restriction-Site Associated DNA Sequencing (RAD-Seq): RAD-Seq uses restriction enzymes to selectively sequence DNA fragments adjacent to recognition sites. It's particularly useful for detecting Single Nucleotide Polymorphisms (SNPs) [102].

b. Genotyping by Sequencing (GBS): GBS involves digesting genomic DNA with restriction enzymes and sequencing the resulting fragments. It has been used to discover SNPs and other genetic variants in various crop species [103].

c. Specific-Locus Amplified Fragment Sequencing (SLAF-Seq): SLAF-Seq combines reduced representation libraries with NGS to identify and genotype SNPs [104].

9.4 Single Nucleotide Polymorphism (SNP)-based platforms

a. Illumina Infinium BeadChip Arrays: these arrays contain thousands to millions of immobilized SNP probes, enabling high-throughput SNP genotyping. They have been extensively used in various crop species [105].

b. Kompetitive Allele Specific PCR (KASP): KASP is a flexible SNP genotyping platform that uses allele-specific primers to target specific SNP loci. It's known for its scalability and cost-effectiveness [106].

9.5 CRISPR-Cas and genotyping

The CRISPR-Cas system has revolutionized genome editing. Researchers have used this technology to develop genotyping methods that exploit its precision and efficiency [107].

10. Genotyping by Sequencing (GBS) and Next Generation Sequencing (NGS) in plant breeding

10.1 Genotyping-by-Sequencing (GBS) in plant breeding

Genotyping-by-Sequencing (GBS) is a high-throughput genotyping method that has gained popularity in plant breeding and genetics. GBS involves the selective sequencing of a subset of the genome, focusing on specific genetic markers, to provide information about genetic variations within a population. This method allows researchers to genotype a large number of samples simultaneously, making it well-suited for plant breeding applications [108].

Here's how GBS works:

1. Library preparation: DNA is extracted from the plant samples of interest. The DNA is then digested using restriction enzymes to create DNA fragments. Unique DNA barcodes (indexes) are added to each sample to allow for multiplexing, where multiple samples are sequenced in the same run.

2. Fragment selection: after digestion, the DNA fragments are size-selected to target specific regions of the genome. This selection can focus on specific genetic markers, such as Single Nucleotide Polymorphisms (SNPs), that are known to be associated with traits of interest.

3. Amplification and sequencing: the selected DNA fragments are amplified using PCR (polymerase chain reaction), and then the amplified fragments are sequenced using next-generation sequencing (NGS) technologies.

4. Data analysis: the sequencing data is processed to identify the genetic variations in the selected regions. By comparing the sequences to a reference genome, researchers can determine the presence of specific alleles and variants associated with traits.

10.2 Next-Generation Sequencing (NGS) technologies in plant breeding

Next-Generation Sequencing (NGS), also known as high-throughput sequencing, refers to a set of modern sequencing technologies that allow rapid and cost-effective sequencing of DNA or RNA molecules. NGS has revolutionized the field of plant breeding by enabling comprehensive analysis of genetic diversity, marker discovery, and trait association studies [109]. Here are some NGS technologies commonly used in plant breeding:

1. Illumina sequencing: illumina platforms, such as HiSeq and NovaSeq, use reversible dye-terminator chemistry to sequence DNA. These platforms offer high-throughput, accurate, and cost-effective sequencing, making them widely used in plant genomics. They have been used for GBS as well as whole-genome sequencing in various plant species.

2. Ion torrent sequencing: ion torrent sequencing is based on the detection of hydrogen ions released during DNA synthesis. This technology is known for its simplicity and speed, making it suitable for targeted sequencing and genotyping applications.

3. Pacific Biosciences (PacBio): PacBio platforms use single-molecule real-time (SMRT) sequencing to generate longer reads compared to Illumina. This longer read length can be valuable for resolving complex regions of the genome, such as repeat-rich areas.

4. Oxford Nanopore Technologies (ONT): ONT's nanopore sequencing technology passes DNA strands through nanopores, and the changes in electrical current as DNA passes through are used to identify the sequence. ONT offers long read lengths and has been used for the de novo assembly of plant genomes.

10.3 Examples of applications

1. Trait mapping: researchers have used GBS to identify genetic markers associated with desirable traits in crops. For example, GBS has been used to map genes responsible for disease resistance, yield, and nutritional content in various plant species.

2. Diversity analysis: NGS technologies have enabled a comprehensive analysis of genetic diversity within crop populations. This information is crucial for selecting diverse parents for breeding programs, conserving germplasm, and understanding the genetic basis of adaptability.

3. Marker-assisted selection (MAS): GBS-derived markers have been integrated into breeding programs for the efficient selection of individuals carrying desired traits. This accelerates the breeding process by allowing early identification of superior genotypes.

4. Genome-wide association studies (GWAS): NGS data, including GBS, have been used in GWAS to identify genomic regions associated with complex traits. This helps breeders understand the genetic basis of trait variation and potentially predict performance.

5. Genome sequencing and annotation: NGS technologies have facilitated the sequencing and annotation of complete plant genomes. This knowledge enhances our understanding of gene function, evolution, and diversity.

6. Molecular breeding: Both GBS and NGS have enabled the development of molecular breeding strategies that exploit genetic information to improve crop varieties for specific traits, leading to faster and more targeted breeding processes.

These examples showcase the power of GBS and NGS technologies in advancing plant breeding research and providing valuable insights into the genetic basis of plant traits and diversity. As technology continues to evolve, these methods will likely play an even more significant role in shaping the future of plant breeding.

11. Integration of molecular marker technology with conventional breeding methods

The integration of molecular marker technology with conventional breeding methods has revolutionized the field of plant breeding, leading to more efficient and targeted breeding strategies. This approach is commonly referred to as "marker-assisted selection" (MAS) or "marker-assisted breeding" (MAB). It involves the use of molecular markers, which are specific DNA sequences associated with particular traits, to aid in the selection of desired traits in plants.

Here's how the integration works and the benefits it brings to plant breeding:

1. Marker identification: first, researchers identify molecular markers that are closely linked to the target traits of interest, such as disease resistance, yield potential, quality characteristics, and more. These markers are typically identified through techniques like genetic mapping and sequencing.

2. Marker selection: once the markers associated with the target traits are identified, breeders can use these markers to select plants that carry the desired traits. This eliminates the need to wait for the trait to be visually expressed in the plant's phenotype, which can take several generations through traditional breeding methods.

3. Early screening: molecular markers allow for early screening of plants in their juvenile stages or even at the seedling level. This rapid screening enables breeders to identify and discard plants that do not possess the desired traits, saving time and resources.

4. Trait introgression: molecular markers can aid in transferring specific traits from one plant variety to another. For example, if a wild plant possesses a desired trait (e.g., disease resistance) but lacks other favorable characteristics (e.g., yield potential), molecular markers can help breeders introgress the desired trait into a high-yielding commercial variety while retaining its positive attributes.

5. Pyramiding traits: breeders can use molecular markers to combine multiple desirable traits into a single plant variety through a process called trait pyramiding. This results in plants with enhanced performance and adaptability.

6. Reduced generation time: by using markers to assist in selecting plants with desired traits, breeders can reduce the number of generations required to develop a new variety. This accelerates the breeding process and allows for faster release of improved varieties to farmers.

7. Precision and accuracy: molecular markers provide a more precise and accurate way to select specific traits compared to relying solely on phenotypic observations, which can be influenced by environmental factors. This precision leads to higher success rates in achieving the desired trait combinations.

8. Preservation of genetic diversity: molecular markers help breeders maintain genetic diversity by allowing them to select specific traits while preserving the overall genetic background of the plant variety.

9. Expanding breeding possibilities: molecular marker technology enables the utilization of traits that may be difficult to select through traditional breeding due to their complex inheritance patterns or low heritability.

In summary, the integration of molecular marker technology with conventional breeding methods offers a powerful toolset for plant breeders to develop new and improved crop varieties more efficiently. This approach has contributed to the development of crop varieties that are better suited to challenges such as pests, diseases, changing climates, and consumer preferences.

12. Challenges and limitations of molecular marker technology

Molecular marker technology has revolutionized plant breeding by allowing for a more precise and efficient selection of desirable traits in plants. However, like any technology, it comes with its own set of challenges and limitations. Here are some of the major challenges and limitations of molecular marker technology in plant breeding:

1. Cost: developing and implementing molecular markers can be expensive, particularly for small-scale breeding programs or in developing countries. The initial investment required for equipment, reagents, and personnel training can be a significant barrier.

2. Marker-trait associations: the accuracy of marker-trait associations is crucial for successful marker-assisted selection (MAS). However, not all markers are perfectly linked to the target trait due to factors like genetic recombination and environmental influences. This can result in false positives or negatives, reducing the efficiency of selection.

3. Marker density and coverage: using a limited number of markers may not capture the entire genetic variation associated with a trait, leading to incomplete information and potential bias in selection. High-density marker panels can mitigate this, but they come with higher costs and computational challenges.

4. Trait complexity: many traits of interest in plant breeding are controlled by multiple genes and their interactions. Molecular markers often identify individual genes or loci, making them less effective for complex traits influenced by multiple genetic factors.

5. Linkage disequilibrium: over time, the linkage between molecular markers and target genes can break down due to genetic recombination, reducing the accuracy of MAS. This is particularly relevant when markers are developed in one population and applied to another.

6. Genetic diversity and transferability: molecular markers developed in one plant population may not be directly transferable to other populations due to differences in genetic background and marker polymorphism. This can limit the broad applicability of certain markers.

7. Ethical and regulatory concerns: the use of genetically modified organisms (GMOs) or markers associated with patented genes can raise ethical and regulatory challenges. This can complicate the adoption of marker-assisted breeding strategies.

8. Limited trait information: for many traits, the underlying genetic mechanisms are not fully understood. In such cases, even if markers are available, their effective use in breeding may be limited by incomplete knowledge of the trait's genetic basis.

9. Data analysis and interpretation: managing and analyzing large datasets generated by high-throughput genotyping technologies can be computationally

demanding and require specialized expertise. Misinterpretation of results can lead to erroneous breeding decisions.

10. Public acceptance: some consumers and stakeholders may have concerns about the use of molecular marker technology in breeding due to perceptions of genetic modification or other safety concerns. Public acceptance can influence the adoption of marker-assisted breeding practices.

11. Breeding timeframes: while molecular markers can expedite the breeding process, developing and validating markers, as well as conducting MAS, still require time. The overall breeding process can be constrained by other factors, such as crop growth cycles.

Despite these challenges and limitations, molecular marker technology continues to play a pivotal role in modern plant breeding, enhancing the efficiency and precision of trait selection. Ongoing advancements in genotyping techniques, bioinformatics, and our understanding of plant genetics will likely help address some of these limitations over time.

13. Future perspectives and emerging trends in molecular marker technology

1. High-throughput sequencing and genotyping: advances in high-throughput sequencing technologies have allowed for cost-effective and rapid sequencing of entire genomes. This has led to the development of genotyping-by-sequencing (GBS) approaches, enabling the identification of a large number of molecular markers across the genome. This trend is likely to continue with improvements in sequencing platforms and data analysis techniques.

2. Precision breeding and genome editing: molecular markers play a critical role in precision breeding, where specific genetic traits can be targeted for modification using genome editing techniques like CRISPR-Cas9. Molecular markers are used to track the edited genes and to ensure the absence of unintended modifications. This technology holds promise for creating new crop varieties with improved traits.

3. Phenomics and genomics integration: integrating genomic data with high-throughput phenotyping data is becoming essential in modern plant breeding. This enables breeders to associate specific genetic markers with desirable phenotypic traits, leading to more informed selection and breeding decisions.

4. Multi-omics approaches: the integration of multiple omics datasets, including genomics, transcriptomics, proteomics, and metabolomics, allows for a more comprehensive understanding of the genetic and molecular mechanisms underlying complex traits. This approach can provide deeper insights into plant biology and aid in identifying key regulatory pathways for targeted improvement.

5. Big data and machine learning: the growing volume of genomic and phenotypic data requires sophisticated computational tools for analysis. Machine learning algorithms can help identify patterns, predict trait outcomes, and assist breeders in selecting the best candidates for breeding programs.

6. Marker-assisted selection (MAS) and genomic selection (GS): these approaches continue to gain traction. MAS involves using specific markers linked to known traits of interest, while GS uses a dense set of markers across the genome to predict an individual's breeding value. These techniques enhance the efficiency of selection by reducing the need for time-consuming phenotypic evaluations.

7. Functional genomics and trait discovery: molecular marker technology is being used in conjunction with functional genomics approaches to decipher the roles of specific genes in trait expression. This can accelerate the discovery of genes responsible for important agronomic traits.

8. Disease resistance and stress tolerance: the identification of markers associated with disease resistance and stress tolerance is of significant interest in plant breeding. Developing crops that are resilient to biotic and abiotic stresses is crucial for ensuring food security.

9. Epigenetics and epigenomics: epigenetic modifications can play a significant role in gene expression and phenotypic variation. Integrating epigenetic data into breeding programs can provide insights into transgenerational effects and potentially improve trait stability.

10. Public-private partnerships and open data sharing: collaboration between public institutions, private companies, and researchers is important for advancing molecular marker technology. Open data sharing fosters innovation and accelerates progress in crop improvement.

14. Conclusions

Molecular marker analysis has significantly advanced plant breeding by enabling breeders to efficiently select and manipulate desired traits. The choice of marker type and analysis technique depends on factors like the target species, available resources, and research objectives. Advances in genomics and technology continue to shape the field, with newer marker types and genotyping platforms being developed over time. Marker-assisted backcrossing is a valuable tool in modern plant breeding that facilitates the efficient transfer of specific traits from donor parents to recipient parents. It combines traditional breeding methods with molecular marker technology, allowing breeders to make precise selections and shorten the time required to develop improved plant varieties with desired traits. Marker-assisted selection has transformed the field of plant breeding by enabling breeders to make more informed and targeted selections in developing improved crop varieties. Its applications are vast and continue to expand, ranging from disease resistance to quality improvement, drought tolerance to yield enhancement, and much more. This technology has played a significant role in accelerating the development of crops that can address the challenges of a changing world.

High-throughput genotyping platforms have significantly advanced plant breeding efforts by providing researchers with powerful tools to dissect the genetic basis of important traits. Over the years, various researchers have contributed to the development and refinement of these platforms, enabling the acceleration of crop improvement through targeted breeding approaches. As technology continues to evolve, genotyping platforms are likely to become even more sophisticated and accessible, facilitating more efficient and precise plant breeding strategies. The power of GBS and NGS technologies in advancing plant breeding research and providing valuable insights into the genetic basis of plant traits and diversity. As technology continues to evolve, these methods will likely play an even more significant role in shaping the future of plant breeding.

In conclusion, molecular marker technology represents a paradigm shift in the field of plant breeding, offering a powerful set of tools that enhance the efficiency, precision, and effectiveness of crop improvement efforts. By enabling breeders to identify and manipulate desired traits at the molecular level, this technology has ushered in a new era of agriculture, where the challenges of global food security and sustainability can be tackled with greater agility and innovation. As our understanding of plant genomics deepens and molecular techniques continue to evolve, the potential for developing novel and improved plant varieties becomes increasingly promising.

Conflict of interest

The authors declare no conflict of interest.

Author details

Vijay Kamal Meena[1*], Harsh Vardhan Singh Shekhawat[1], Subhash Chand[2], Kapil Choudhary[3], Jitendra Kumar Sharma[1] and Lekha Lekha[1]

1 Agricultural Research Sub-Station (Sumerpur), Agriculture University, Jodhpur, India

2 ICAR—Indian Grassland and Fodder Research Institute, Jhansi, India

3 College of Agriculture (Sumerpur), Agriculture University, Jodhpur, India

*Address all correspondence to: vjkamal93@gmail.com

References

[1] Nadeem MA, Nawaz MA, Shahid MQ, Doğan Y, Comertpay G, Yıldız M, et al. DNA molecular markers in plant breeding: Current status and recent advancements in genomic selection and genome editing. Biotechnology & Biotechnological Equipment. 2018;**32**:261-285. DOI: 10.1080/13102818.2017.1400401

[2] Soriano JM. Molecular marker technology for crop improvement. Agronomy. 2020;**10**:1462. DOI: 10.3390/agronomy10101462

[3] Winter P, Kahl G. Molecular marker technologies for plant improvement. World Journal of Microbiology and Biotechnology. 1995;**11**:438-448. DOI: 10.1007/BF00364619

[4] Bohar R, Chitkineni A, Varshney RK. Genetic molecular markers to accelerate genetic gains in crops. BioTechniques. 2020;**69**:158-160. DOI: 10.2144/btn-2020-0066

[5] Nair RJ, Pandey MK. Role of molecular markers in crop breeding: A review. Agricultural Reviews. 2021;**1**:1-8. DOI: 10.18805/ag.R-2322

[6] Ahmar S, Gill RA, Jung K-H, Faheem A, Qasim MU, Mubeen M, et al. Conventional and molecular techniques from simple breeding to speed breeding in crop plants: Recent advances and future outlook. International Journal of Molecular Sciences. 2020;**21**:2590. DOI: 10.3390/ijms21072590

[7] Glenn KC, Alsop B, Bell E, Goley M, Jenkinson J, Liu B, et al. Bringing new plant varieties to market: Plant breeding and selection practices advance beneficial characteristics while minimizing unintended changes. Crop Science. 2017;**57**:2906-2921. DOI: 10.2135/cropsci2017.03.0199

[8] Lema M. Marker assisted selection in comparison to conventional plant breeding: Review article. Agricultural Research & Technology. 2018;**14**:555914. DOI: 10.19080/ARTOAJ.2018.14.555914

[9] Meena VK, Taak Y, Chaudhary R, Chand S, Patel MK, Muthusamy V, et al. Deciphering the genetic inheritance of tocopherols in Indian Mustard (*Brassica Juncea* L. Czern and Coss). Plants. 2022;**11**:1779. DOI: 10.3390/plants11131779

[10] Hasan N, Choudhary S, Naaz N, Sharma N, Laskar RA. Recent advancements in molecular marker-assisted selection and applications in plant breeding programmes. Journal of Genetic Engineering and Biotechnology. 2021;**19**:128. DOI: 10.1186/s43141-021-00231-1

[11] Jiang G-L. Molecular marker-assisted breeding: A plant Breeder's review. In: Advances in Plant Breeding Strategies: Breeding, Biotechnology and Molecular Tools. Cham: Springer International Publishing; 2015. pp. 431-472

[12] Appleby N, Edwards D, Batley J. New technologies for ultra-high throughput genotyping in plants. In: Gustafson JP, Langridge P, Somers DJ, Totowa NJ, editors. Methods in Molecular Biology, Plant Genomics. New York: Humana Press; 2009. p. 19-39. DOI: 10.1007/978-1-59745-427-8_2

[13] MirMohammadi Maibody SAM, Golkar P. Application of DNA molecular markers in plant breeding (review article). Plant Genetic Researches. 2019;**6**:1-30. DOI: 10.29252/pgr.6.1.1

[14] Landry BS, Michelmore RW. Methods and applications of restriction fragment length polymorphism analysis to plants. In: Tailoring Genes for Crop Improvement. Boston, MA: Springer US; 1987. pp. 25-44

[15] Williams RC. Restriction fragment length polymorphism (RFLP). American Journal of Physical Anthropology. 1989;**32**:159-184. DOI: 10.1002/ajpa.1330320508

[16] Blears MJ, De Grandis SA, Lee H, Trevors JT. Amplified fragment length polymorphism (AFLP): A review of the procedure and its applications. Journal of Industrial Microbiology & Biotechnology. 1998;**21**:99-114. DOI: 10.1038/sj.jim.2900537

[17] Simpson J. Amplified fragment length polymorphisms (AFLP's). Botanical Sciences. 1997;**60**:119-122. DOI: 10.17129/botsci.1524

[18] Godwin ID, Aitken EAB, Smith LW. Application of inter simple sequence repeat (ISSR) markers to plant genetics. Electrophoresis. 1997;**18**:1524-1528. DOI: 10.1002/elps.1150180906

[19] Kalia RK, Rai MK, Kalia S, Singh R, Dhawan AK. Microsatellite markers: An overview of the recent progress in plants. Euphytica. 2011;**177**:309-334. DOI: 10.1007/s10681-010-0286-9

[20] Meena VK, Sharma RK, Chand S, Kumar M, Kumar N, Jain N, et al. Elucidating molecular diversity in spring wheat (*Triticum Aestivum* L. Em. Thell.) under terminal heat stress environment using morpho-physiological traits and SSR markers. Indian Journal of Genetics and Plant Breeding. 2022;**82**:47-55. DOI: 10.31742/IJGPB.82.1.7

[21] Batra J, Srinivasan S, Clements J. Single nucleotide polymorphisms (SNPs). In: Molecular Testing in Cancer. New York: New York, NY: Springer; 2014. pp. 55-80

[22] Marth GT, Korf I, Yandell MD, Yeh RT, Gu Z, Zakeri H, et al. A general approach to single-nucleotide polymorphism discovery. Nature Genetics. 1999;**23**:452-456. DOI: 10.1038/70570

[23] Kaundun S, Matsumoto S. Development of CAPS markers based on three key genes of the phenylpropanoid pathway in tea, Camellia sinensis (L.) O. Kuntze, and differentiation between Assamica and sinensis varieties. Theoretical and Applied Genetics. 2003;**106**:375-383. DOI: 10.1007/s00122-002-0999-9

[24] Neff MM, Neff JD, Chory J, Pepper AE. DCAPS, a simple technique for the genetic analysis of single nucleotide polymorphisms: Experimental applications in *Arabidopsis thaliana* genetics. The Plant Journal. 1998;**14**:387-392. DOI: 10.1046/j.1365-313X.1998.00124.x

[25] Atienzar FA, Jha AN. The random amplified polymorphic DNA (RAPD) assay and related techniques applied to genotoxicity and carcinogenesis studies: A critical review. Mutation Research/Reviews in Mutation Research. 2006;**613**:76-102. DOI: 10.1016/j.mrrev.2006.06.001

[26] Williams JGK, Kubelik AR, Livak KJ, Rafalski JA, Tingey SV. DNA polymorphisms amplified by arbitrary primers are useful as genetic markers. Nucleic Acids Research. 1990;**18**:6531-6535. DOI: 10.1093/nar/18.22.6531

[27] Guasmi F, Elfalleh W, Hannachi H, Fères K, Touil L, Marzougui N, et al. The use of ISSR and RAPD markers for genetic diversity among south Tunisian

barley. ISRN Agronomy. 2012;**2012**:1-10. DOI: 10.5402/2012/952196

[28] Kantety RV, Zeng X, Bennetzen JL, Zehr BE. Assessment of genetic diversity in dent and popcorn (*Zea mays* L.) inbred lines using inter-simple sequence repeat (ISSR) amplification. Molecular Breeding. 1995;**1**:365-373. DOI: 10.1007/BF01248414

[29] Chen H, Xie W, He H, Yu H, Chen W, Li J, et al. A high-density SNP genotyping array for rice biology and molecular breeding. Molecular Plant. 2014;**7**:541-553. DOI: 10.1093/mp/sst135

[30] Ganal MW, Polley A, Graner E-M, Plieske J, Wieseke R, Luerssen H, et al. Large SNP arrays for genotyping in crop plants. Journal of Biosciences. 2012;**37**:821-828. DOI: 10.1007/s12038-012-9225-3

[31] Elshire RJ, Glaubitz JC, Sun Q, Poland JA, Kawamoto K, Buckler ES, et al. A robust, simple genotyping-by-sequencing (GBS) approach for high diversity species. PLoS One. 2011;**6**:e19379. DOI: 10.1371/journal.pone.0019379

[32] Fu Y-B, Dong Y. PaSNPg: A GBS-based pipeline for protein-associated SNP discovery and genotyping in non-model species. Journal of Proteomics & Bioinformatics. 2015;**08**:190-194. DOI: 10.4172/jpb.1000368

[33] Bird A. DNA methylation patterns and epigenetic memory. Genes & Development. 2002;**16**:6-21. DOI: 10.1101/gad.947102

[34] Laird PW. Principles and challenges of genome-wide DNA methylation analysis. Nature Reviews. Genetics. 2010;**11**:191-203. DOI: 10.1038/nrg2732

[35] Gupta PK, Rustgi S. Molecular markers from the transcribed/expressed region of the genome in higher plants. Functional & Integrative Genomics. 2004;**4**. DOI: 10.1007/s10142-004-0107-0

[36] Harbers M, Carninci P. Tag-based approaches for transcriptome research and genome annotation. Nature Methods. 2005;**2**:495-502. DOI: 10.1038/nmeth768

[37] Parla JS, Iossifov I, Grabill I, Spector MS, Kramer M, McCombie WR. A comparative analysis of exome capture. Genome Biology. 2011;**12**:R97. DOI: 10.1186/gb-2011-12-9-r97

[38] Zhao M, Wang Q, Wang Q, Jia P, Zhao Z. Computational tools for copy number variation (CNV) detection using next-generation sequencing data: Features and perspectives. BMC Bioinformatics. 2013;**14**:S1. DOI: 10.1186/1471-2105-14-S11-S1

[39] Wang C, Li L, Zhang X, Gao Q, Wang R, An D. Development and application of EST-STS markers specific to chromosome 1RS of secale cereale. Cereal Research Communications. 2009;**37**:13-21. DOI: 10.1556/CRC.37.2009.1.2

[40] Adhikari S, Saha S, Biswas A, Rana TS, Bandyopadhyay TK, Ghosh P. Application of molecular markers in plant genome analysis: A review. The Nucleus. 2017;**60**:283-297. DOI: 10.1007/s13237-017-0214-7

[41] Jiang G-L. Molecular markers and marker-assisted breeding in plants. In: Plant Breeding from Laboratories to Fields. London, UK: InTech; 2013

[42] Cui Y, Zhang F, Xu J, Li Z, Xu S. Mapping quantitative trait loci in selected breeding populations:

A segregation distortion approach. Heredity (Edinb). 2015;**115**:538-546. DOI: 10.1038/hdy.2015.56

[43] van Eeuwijk FA, Bink MC, Chenu K, Chapman SC. Detection and use of QTL for complex traits in multiple environments. Current Opinion in Plant Biology. 2010;**13**:193-205. DOI: 10.1016/j.pbi.2010.01.001

[44] Schulman AH. Molecular markers to assess genetic diversity. Euphytica. 2007;**158**:313-321. DOI: 10.1007/s10681-006-9282-5

[45] Yang H, Li C, Lam H-M, Clements J, Yan G, Zhao S. Sequencing consolidates molecular markers with plant breeding practice. Theoretical and Applied Genetics. 2015;**128**:779-795. DOI: 10.1007/s00122-015-2499-8

[46] Kidwell KK, Osborn TC. Simple plant DNA isolation procedures. In: Plant Genomes: Methods for Genetic and Physical Mapping. Springer Netherlands: Dordrecht; 1992. pp. 1-13

[47] Wojdacz TK, Dobrovic A. Methylation-sensitive high-resolution melting (MS-HRM): A new approach for sensitive and high-throughput assessment of methylation. Nucleic Acids Research. 2007;**35**:e41-e41. DOI: 10.1093/nar/gkm013

[48] Metzker ML. Emerging technologies in DNA sequencing. Genome Research. 2005;**15**:1767-1776. DOI: 10.1101/gr.3770505

[49] Thomson MJ. High-throughput SNP genotyping to accelerate crop improvement. Plant Breeding and Biotechnology. 2014;**2**:195-212. DOI: 10.9787/PBB.2014.2.3.195

[50] Xu Y, Crouch JH. Marker-assisted selection in plant breeding: From publications to practice. Crop Science. 2008;**48**:391-407. DOI: 10.2135/cropsci2007.04.0191

[51] Kendziorski C, Wang P. A review of statistical methods for expression quantitative trait loci mapping. Mammalian Genome. 2006;**17**:509-517. DOI: 10.1007/s00335-005-0189-6

[52] Alseekh S, Kostova D, Bulut M, Fernie AR. Genome-wide association studies: Assessing trait characteristics in model and crop plants. Cellular and Molecular Life Sciences. 2021;**78**:5743-5754. DOI: 10.1007/s00018-021-03868-w

[53] Collard BCY, Jahufer MZZ, Brouwer JB, Pang ECK. An introduction to markers, quantitative trait loci (QTL) mapping and marker-assisted selection for crop improvement: The basic concepts. Euphytica. 2005;**142**:169-196. DOI: 10.1007/s10681-005-1681-5

[54] Nguyen HT, Vu QH, Van Mai T, Nguyen TT, Vu LD, Nguyen TT, et al. Marker-assisted selection of Xa21 conferring resistance to bacterial leaf blight in indica Rice cultivar LT2. Rice Science. 2018;**25**:52-56. DOI: 10.1016/j.rsci.2017.08.004

[55] Mahajan S, Tuteja N. Cold, salinity and drought stresses: An overview. Archives of Biochemistry and Biophysics. 2005;**444**:139-158. DOI: 10.1016/j.abb.2005.10.018

[56] Cominelli E, Conti L, Tonelli C, Galbiati M. Challenges and perspectives to improve crop drought and salinity tolerance. New Biotechnology. 2013;**30**:355-361. DOI: 10.1016/j.nbt.2012.11.001

[57] Mondini L, Noorani A, Pagnotta M. Assessing plant genetic diversity by molecular tools. Diversity (Basel). 2009;**1**:19-35. DOI: 10.3390/d1010019

[58] Wang F, Tian H, Yi H, Zhao H, Huo Y, Kuang M, et al. Principle and strategy of DNA fingerprint identification of plant variety. Molecular Plant Breeding. 2019;**10**:81-92. DOI: 10.5376/mpb.2019.10.0011

[59] Li Q, Zhang T, Li L, Bao Z, Tu W, Xiang P, et al. Comparative mitogenomic analysis reveals intraspecific, interspecific variations and genetic diversity of medical fungus ganoderma. Journal of Fungi. 2022;**8**:781. DOI: 10.3390/jof8080781

[60] Zhang J, Franks RG, Liu X, Kang M, Keebler JEM, Schaff JE, et al. De novo sequencing, characterization, and comparison of inflorescence transcriptomes of *Cornus Canadensis* and *C. Florida* (Cornaceae). PLoS One. 2013;**8**:e82674. DOI: 10.1371/journal.pone.0082674

[61] Liu S, An Y, Tong W, Qin X, Samarina L, Guo R, et al. Characterization of genome-wide genetic variations between two varieties of tea plant (*Camellia sinensis*) and development of InDel markers for genetic research. BMC Genomics. 2019;**20**:935. DOI: 10.1186/s12864-019-6347-0

[62] An H, Lee H-Y, Shim D, Choi SH, Cho H, Hyun TK, et al. Development of CAPS markers for evaluation of genetic diversity and population structure in the germplasm of button mushroom (*Agaricus bisporus*). Journal of Fungi. 2021;7:375. DOI: 10.3390/jof7050375

[63] Kumekawa Y, Kilmaskossu M, Mori M, Miyazaki A, Ito K, Arakawa R, et al. Changes in plant species during succession in a sago Forest. American Journal of Plant Sciences. 2014;**05**:3526-3534. DOI: 10.4236/ajps.2014.524369

[64] Nerva L, Dalla Costa L, Ciacciulli A, Sabbadini S, Pavese V, Dondini L, et al. The role of Italy in the use of advanced plant genomic techniques on fruit trees: State of the art and future perspectives. International Journal of Molecular Sciences. 2023;**24**:977. DOI: 10.3390/ijms24020977

[65] Zainal-Abidin R-A, Ruhaizat-Ooi I-H, Harun S. A review of omics technologies and bioinformatics to accelerate improvement of papaya traits. Agronomy. 2021;**11**:1356. DOI: 10.3390/agronomy11071356

[66] Wu J-G, Shi C-H, Zhang H-Z. Partitioning genetic effects due to embryo, cytoplasm and maternal parent for oil content in oilseed rape (*Brassica napus* L.). Genetics and Molecular Biology. 2006;**29**:533-538. DOI: 10.1590/S1415-47572006000300023

[67] Elattar MA, Karikari B, Li S, Song S, Cao Y, Aslam M, et al. Identification and validation of major QTLs, epistatic interactions, and candidate genes for soybean seed shape and weight using two related RIL populations. Frontiers in Genetics. 2021;**12**:666440. DOI: 10.3389/fgene.2021.666440

[68] Chhuneja P, Kaur S, Dhaliwal HS. Introgression and exploitation of biotic stress tolerance from related wild species in wheat cultivars. In: Rajpal V, Rao S, Raina S, editors. Molecular Breeding for Sustainable Crop Improvement. Sustainable Development and Biodiversity. Vol. 11. Cham: Springer; 2017. pp. 269-324. DOI: 10.1007/978-3-319-27090-6_12

[69] Randhawa MS, Bains NS, Sohu VS, Chhuneja P, Trethowan RM, Bariana HS, et al. Marker assisted transfer of stripe rust and stem rust resistance genes into four wheat cultivars. Agronomy. 2019;**9**:497. DOI: 10.3390/agronomy9090497

[70] Jannink J-L, Lorenz AJ, Iwata H. Genomic selection in plant breeding: From theory to practice. Briefings in Functional Genomics. 2010;**9**:166-177. DOI: 10.1093/BGP/elq001

[71] Crossa J, Pérez P, de los Campos G, Mahuku G, Dreisigacker S, Magorokosho C. Genomic selection and prediction in plant breeding. Journal of Crop Improvement. 2011;**25**:239-261. DOI: 10.1080/15427528.2011.558767

[72] Crossa J, Campos G d l, Pérez P, Gianola D, Burgueño J, Araus JL, et al. Prediction of genetic values of quantitative traits in plant breeding using pedigree and molecular markers. Genetics. 2010;**186**:713-724. DOI: 10.1534/genetics.110.118521

[73] Ahmad R, Anjum MA. Applications of molecular markers to assess genetic diversity in vegetable and ornamental crops - A review. Journal of Horticultural Science & Technology. 2018;**1**:1-7. DOI: 10.46653/jhst180101001

[74] Goswami M, Attri K, Goswami I. Applications of molecular markers in fruit crops: A review. International Journal of Economic Plants. 2022;**9**:121-126. DOI: 10.23910/2/2022.0459

[75] Lahogue F, This P, Bouquet A. Identification of a codominant scar marker linked to the seedlessness character in grapevine. Theoretical and Applied Genetics. 1998;**97**:950-959. DOI: 10.1007/s001220050976

[76] Zhang H, Fan X, Zhang Y, Jiang J, Liu C. Identification of Favorable SNP alleles and candidate genes for seedlessness in *Vitis vinifera* L. Using Genome-Wide Association Mapping. Euphytica. 2017;**213**:136. DOI: 10.1007/s10681-017-1919-z

[77] Tomizawa E, Ohtomo S, Asai K, Ohta Y, Takiue Y, Hasumi A, et al. Additional betalain accumulation by genetic engineering leads to a novel flower color in lisianthus (*Eustoma grandiflorum*). Plant Biotechnology. 2021;**38**:21.0516a. DOI: 10.5511/plantbiotechnology.21.0516a

[78] Agarwal M, Shrivastava N, Padh H. Advances in molecular marker techniques and their applications in plant sciences. Plant Cell Reports. 2008;**27**:617-631. DOI: 10.1007/s00299-008-0507-z

[79] Singh M, Randhawa G, Bhoge RK, Singh S, Kak A, Sangwan O. Monitoring adventitious presence of transgenes in cotton collections from Genebank and experimental plots: Ensuring GM-free conservation and cultivation of genetic resources. Agricultural Research. 2020;**9**:469-476. DOI: 10.1007/s40003-019-00449-z

[80] Zargar SM, Raatz B, Sonah H, Nazir M, Bhat JA, Dar ZA, et al. Recent advances in molecular marker techniques: Insight into QTL mapping, GWAS and genomic selection in plants. Journal of Crop Science and Biotechnology. 2015;**18**:293-308. DOI: 10.1007/s12892-015-0037-5

[81] Arús P, Moreno-González J. Marker-assisted selection. In: Plant Breeding. Springer Netherlands: Dordrecht; 1993. pp. 314-331

[82] Muhammad Gul Arabzai. Hameed gul application techniques of molecular marker and achievement of marker assisted selection (MAS) in three major crops Rice, wheat and maize. International Journal for Research in Applied Sciences and Biotechnology. 2021;**8**:82-93. DOI: 10.31033/ijrasb.8.1.10

[83] Ortega F, Lopez-Vizcon C. Application of molecular marker-assisted

selection (MAS) for disease resistance in a practical potato breeding programme. Potato Research. 2012;**55**:1-13. DOI: 10.1007/s11540-011-9202-5

[84] Song L, Wang R, Yang X, Zhang A, Liu D. Molecular markers and their applications in marker-assisted selection (MAS) in bread wheat (*Triticum aestivum* L.). Agriculture. 2023;**13**:642. DOI: 10.3390/agriculture13030642

[85] Cattivelli L, Rizza F, Badeck F-W, Mazzucotelli E, Mastrangelo AM, Francia E, et al. Drought tolerance improvement in crop plants: An integrated view from breeding to genomics. Field Crops Research. 2008;**105**:1-14. DOI: 10.1016/j.fcr.2007.07.004

[86] Krannich C, Maletzki L, Kurowsky C, Horn R. Network candidate genes in breeding for drought tolerant crops. International Journal of Molecular Sciences. 2015;**16**:16378-16400. DOI: 10.3390/ijms160716378

[87] Ferrante A, Nocito FF, Morgutti S, Sacchi GA. Plant breeding for improving nutrient uptake and utilization efficiency. In: Tei F, Nicola S, Benincasa P, editors. Plant Breeding for Improving Nutrient Uptake and Utilization Efficiency. Advances in Research on Fertilization Management of Vegetable Crops. Advances in Olericulture. Cham: Springer; 2017. p. 221-246. DOI: 10.1007/978-3-319-53626-2_8

[88] Osei MK, Prempeh R, Adjebeng-Danquah JA, Opoku J, Danquah A, Danquah E, et al. Marker-assisted selection (MAS): A fast-track tool in tomato breeding. In: Recent Advances in Tomato Breeding and Production. London, UK, London, UK: IntechOpen; 2019

[89] Dong H, Huang Y, Wang K. The development of herbicide resistance crop plants using CRISPR/Cas9-mediated gene editing. Genes (Basel). 2021;**12**:912. DOI: 10.3390/genes12060912

[90] Heffner EL, Sorrells ME, Jannink J-L. Genomic selection for crop improvement. Crop Science. 2009;**49**: 1-12. DOI: 10.2135/cropsci2008.08.0512

[91] Shikha M, Kanika A, Rao AR, Mallikarjuna MG, Gupta HS, Nepolean T. Genomic selection for drought tolerance using genome-wide SNPs in maize. Frontiers in Plant Science. 2017;**8**:00550. DOI: 10.3389/fpls.2017.00550

[92] Yao J, Zhao D, Chen X, Zhang Y, Wang J. Use of genomic selection and breeding simulation in cross prediction for improvement of yield and quality in wheat (*Triticum aestivum* L.). Crop Journal. 2018;**6**:353-365. DOI: 10.1016/j.cj.2018.05.003

[93] Ahmadi N, Bartholomé J, Cao T-V, Grenier C. Genomic selection in rice: Empirical results and implications for breeding. In: Quantitative Genetics, Genomics and Plant Breeding. UK: CABI; 2020. pp. 243-258

[94] Matei G, Woyann LG, Milioli AS, de Bem Oliveira I, Zdziarski AD, Zanella R, et al. Genomic selection in soybean: Accuracy and time gain about phenotypic selection. Molecular Breeding. 2018;**38**:117. DOI: 10.1007/s11032-018-0872-4

[95] Kumar S, Chagné D, Bink MCAM, Volz RK, Whitworth C, Carlisle C. Genomic selection for fruit quality traits in apple (Malus × domestica Borkh.). PLoS One. 2012;**7**:36674. DOI: 10.1371/journal.pone.0036674

[96] Frisch M, Melchinger AE. Marker-assisted backcrossing for introgression of a recessive gene. Crop Science. 2001;**41**:1485-1494. DOI: 10.2135/cropsci2001.4151485x

[97] Kadirvel P, Senthilvel S, Geethanjali S, Sujatha M, Varaprasad KS. Genetic markers, trait mapping and marker-assisted selection in plant breeding. In: Bahadur B, Venkat Rajam M, Sahijram L, Krishnamurthy K, editors. Plant Biology and Biotechnology. New Delhi: Springer; 2015. p. 65-88. DOI: 10.1007/978-81-322-2283-5_4

[98] Amom T, Nongdam P. The use of molecular marker methods in plants: A review. International Journal of Current Research and Review. 2017;**9**:1-7. DOI: 10.7324/IJCRR.2017.9171

[99] Deschamps S, Llaca V, May GD. Genotyping-by-sequencing in plants. Biology (Basel). 2012;**1**:460-483. DOI: 10.3390/biology1030460

[100] Reyes VP, Kitony JK, Nishiuchi S, Makihara D, Doi K. Utilization of genotyping-by-sequencing (GBS) for rice pre-breeding and improvement: A review. Life. 2022;**12**:1752. DOI: 10.3390/life12111752

[101] Risterucci A-M, Hippolyte I, Perrier X, Xia L, Caig V, Evers M, et al. Development and assessment of diversity arrays technology for high-throughput DNA analyses in Musa. Theoretical and Applied Genetics. 2009;**119**:1093-1103. DOI: 10.1007/s00122-009-1111-5

[102] Zhai Z, Zhao W, He C, Yang K, Tang L, Liu S, et al. SNP discovery and genotyping using restriction-site-associated DNA sequencing in chickens. Animal Genetics. 2015;**46**:216-219. DOI: 10.1111/age.12250

[103] Spindel J, Wright M, Chen C, Cobb J, Gage J, Harrington S, et al. Bridging the genotyping gap: Using genotyping by sequencing (GBS) to add high-density SNP markers and new value to traditional bi-parental mapping and breeding populations. Theoretical and Applied Genetics. 2013;**126**:2699-2716. DOI: 10.1007/s00122-013-2166-x

[104] Sun X, Liu D, Zhang X, Li W, Liu H, Hong W, et al. SLAF-Seq: An efficient method of large-scale de novo SNP discovery and genotyping using high-throughput sequencing. PLoS One. 2013;**8**:e58700. DOI: 10.1371/journal.pone.0058700

[105] Ganal MW, Wieseke R, Luerssen H, Durstewitz G, Graner E-M, Plieske J, et al. High-throughput SNP profiling of genetic resources in crop plants using genotyping arrays. In: Genomics of Plant Genetic Resources. Dordrecht: Springer Netherlands; 2014. pp. 113-130

[106] Majeed U, Darwish E, Rehman SU, Zhang X. Kompetitive allele specific PCR (KASP): A Singleplex genotyping platform and its application. Journal of Agricultural Science. 2018;**11**:11. DOI: 10.5539/jas.v11n1p11

[107] Manghwar H, Lindsey K, Zhang X, Jin S. CRISPR/Cas system: Recent advances and future prospects for genome editing. Trends in Plant Science. 2019;**24**:1102-1125. DOI: 10.1016/j.tplants.2019.09.006

[108] Chung YS, Choi SC, Jun T-H, Kim C. Genotyping-by-sequencing: A promising tool for plant genetics research and breeding. Horticulture, Environment and Biotechnology. 2017;**58**:425-431. DOI: 10.1007/s13580-017-0297-8

[109] Hodzic J, Gurbeta L, OmanovicMiklicanin E, Badnjevic A. Overview of next-generation sequencing platforms used in published draft plant genomes in light of genotypization of immortelle plant (*Helichrysium arenarium*). Medical Archives. 2017;**71**:288. DOI: 10.5455/medarh.2017.71.288-292

Chapter 2

Allele Mining and Development of Kompetitive Allele Specific PCR (KASP) Marker in Plant Breeding

Hemant Sharma, Sourabh Kumar and Deepa Bhadana

Abstract

Crop improvement refers to the systematic approach of discovering and selecting plants that possess advantageous alleles for specific target genes. The foundation of crop improvement initiatives typically relies on the fundamental concepts of genetic diversity and the genetic architecture of agricultural plants. Allele mining is a contemporary and efficacious technique utilized for the identification of naturally occurring allelic variations within genes that exhibit advantageous characteristics. Consequently, the utilization of allele mining has significant potential as a feasible approach for enhancing crop-related endeavors. The gene pool of a plant exhibits a substantial degree of genetic variety, characterized by the presence of a multitude of mechanism genes. The utilization of genetic variants for the detection and separation of novel alleles of genes that display favorable traits from the current gene pool, and their subsequent incorporation into the development of improved cultivars through the application of marker-assisted selection, is of utmost importance.

Keywords: allele, PCR, genotyping, marker, abiotic and biotic stress

1. Introduction

Plant breeding is a vital field in agriculture that focuses on generating crop varieties with enhanced features to suit the expanding worldwide demand for food, feed, fiber, and bioenergy. Over the decades, conventional breeding methods have played a key role in crop improvement [1, 2]. However, the advent of molecular biology and genomics has revolutionized the profession, giving plant breeders with sophisticated tools to examine the genetic basis of desired traits. One of these technologies, known as allele mining, has emerged as a basic approach to detect and harness beneficial genetic variants within plant populations. In this comprehensive discussion, we will study the notion of allele mining, its significance in identifying favorable genetic variations, its usefulness in boosting agricultural attributes, and the methodologies and procedures employed in this process [3].

The process of introducing desirable genetic variants or alleles into the genetic makeup of a plant population or cultivar is referred to as allele introduction in plants [4]. This can be accomplished through a variety of breeding approaches that aim to increase individual features or overall performance. Allele introduction is an important stage in plant breeding because it allows breeders to harness genetic variation and develop new plant types with desirable traits. The practice of searching and detecting novel genetic variations or alleles within a species' gene pool is referred to as allele mining in plant breeding [5]. This strategy seeks to identify and exploit valuable genetic resources for agricultural improvement. Breeders can acquire desirable features that may be absent or underrepresented in cultivated varieties by tapping into existing genetic diversity. Breeders can use these techniques to transfer desirable alleles into plants and generate new varieties with improved features including disease resistance, stress tolerance, yield potential, qualitative qualities, or agronomic performance. The technique of choice is determined by the specific breeding objectives, the availability of genetic resources, and the features of the target plant species. KASP markers, also known as Kompetitive allele-specific PCR markers, are a type of molecular marker that find widespread application in the field of genetic analysis and plant breeding. They operate on the premise of allele-specific amplification through the utilization of real-time PCR that is based on fluorescence [6]. For the purpose of conducting large-scale screenings of the genetic variants that exist within plant populations, KASP markers offer a solution for genotyping that is high-throughput, quick, and economical. In the field of plant breeding, KASP markers provide a number of benefits, some of which are their high level of specificity, precision, scalability, and cost-effectiveness. They are very helpful when it comes to genotyping huge populations, carrying out high-throughput screening, and locating genetic variants that are connected with key features. In order to hasten the process of selecting better crop varieties and developing new ones, plant breeding programs have extensively included KASP markers into their procedures.

2. Allele mining and identifying desirable genetic variations

Allele mining, also known as allelic variation finding, is a rigorous and methodical strategy designed to investigate and ascertain the genetic variety present in a particular plant species or population. The fundamental objective of allele mining is the identification of alleles, which are variant forms of genes, that exhibit associations with specific qualities of interest. The alleles can exhibit variances in their DNA sequences, which can subsequently lead to differences in the expression or functionality of the corresponding genes [7]. The process of allele mining is of great significance in the exploration of the latent genetic capacity of crops, as it enables the identification of genetic variations that can be utilized for the enhancement of agricultural traits. The initial step in the process of allele mining is the acquisition of genetic data from a wide range of plant individuals or groups belonging to a certain species [8]. The variety in question encompasses a range of plant materials, such as conventional landraces, wild cousins, mutant lines, and other specimens that possess genetic variants with possible implications. The genetic information is thereafter submitted to meticulous study, frequently employing sophisticated molecular biology techniques and genomic tools [3, 9, 10].

3. Enhancing crop traits through allele mining

The significance of allele mining in plant breeding is broad and involves various fundamental areas of crop enhancement:

a. Increased Crop Yields: One of the main goals of plant breeding is to enhance crop varieties that can achieve higher yields in order to address the increasing worldwide demand for food and agricultural commodities. The process of allele mining plays a crucial role in the pursuit of this objective as it enables the identification of alleles that are linked to enhanced yield potential. These alleles have the potential to influence multiple facets of crop development, such as seed size, grain count, and photosynthetic efficiency.

b. Disease Resistance: Crop diseases produced by various pathogens, including bacteria, fungus, viruses, and nematodes, present substantial obstacles to the field of agriculture. The identification of alleles associated with disease resistance is of paramount importance in the development of crops that possess the ability to survive attacks by pathogens. Through the process of allele mining, breeders are able to identify genetic differences that provide inherent resistance to certain conditions. This allows for the development of crops that necessitate fewer chemical treatments, thereby mitigating environmental consequences and decreasing production expenses.

c. Stress Tolerance: Stress tolerance is a critical factor affecting crop output, as environmental stressors such as drought, heat, salt, and soil nutrient deficits can have a significant negative influence. The process of allele mining enables the identification of alleles that are correlated with stress tolerance, so enabling breeders to cultivate crop varieties that exhibit resilience in unfavorable environmental conditions. These genotypes have the potential to impact physiological systems, such as water consumption efficiency or ion transport, which play a crucial role in stress adaption.

d. Nutritional Quality: Nutritional Quality: In conjunction with enhancing crop productivity and stress tolerance, allele mining can also be employed to enhance the nutritional composition of agricultural produce. The identification of genes responsible for the synthesis of crucial minerals, vitamins, and antioxidants has the potential to facilitate the cultivation of crops with heightened nutritional content. This advancement could effectively tackle worldwide health issues associated with malnutrition and dietary insufficiencies.

4. Methods and techniques used for allele mining

The attainment of success in allele mining necessitates the utilization of a synergistic amalgamation of sophisticated methodologies and procedures to investigate and evaluate the genetic variability present within plant populations. Several prominent strategies and techniques utilized in the process of allele mining encompass:

a. Next-Generation Sequencing (NGS): The advent of Next-Generation Sequencing (NGS) has brought about a significant transformation in the field of allele mining, since it allows for the efficient and economical sequencing of complete plant genomes [9, 10]. This methodology, sometimes referred to as whole-genome sequencing, offers a comprehensive perspective on the genetic composition of a plant, enabling the identification of particular genes and alleles linked to advantageous features. Moreover, the utilization of RNA sequencing (RNA-seq) enables researchers to evaluate gene expression patterns, so augmenting our comprehension of how alleles exert an influence on features.

b. Bioinformatics Tools: The utilization of bioinformatics tools is essential for the intricate examination of the extensive genomic datasets produced by Next-Generation Sequencing (NGS) techniques. Bioinformatics tools and software play a crucial role in the processing and interpretation of this data. These methods facilitate the identification of candidate alleles, the prediction of their possible impacts on phenotypes, and the execution of genome-wide association studies (GWAS) to establish links between alleles and phenotypic variants.

c. Marker-Assisted Selection (MAS): Marker-Assisted Selection (MAS) is a technique employed in breeding programs to incorporate possible candidate alleles that have been identified by allele mining. The utilization of molecular markers, such as single nucleotide polymorphisms (SNPs) or simple sequence repeats (SSRs), that are associated with the target alleles, is employed for the purpose of screening and selecting plants that possess these desired genetic variants. The elimination of time-consuming and resource-intensive phenotypic tests expedites the breeding process.

d. Population Genetics and Association Mapping: Population genetics and association mapping are important tools in the field of allele mining, as they offer valuable insights on the genetic diversity and structure of plant populations. Through the examination of the frequency and distribution patterns of particular alleles within populations, scholars are able to discern regions of significance within the genome. Association mapping approaches facilitate the discovery of alleles that are linked to specific features through the analysis of the relationship between genetic markers and differences in phenotypic expression.

e. Functional Genomics: Functional genomics plays a crucial role in not only finding potential alleles but also in unraveling the underlying biological pathways that contribute to the features of interest. Researchers employ several methodologies, including gene expression analyses, gene knockout studies, and gene editing with CRISPR-Cas9 technology, to gain insights into the impact of certain alleles on the molecular mechanisms underlying phenotypic development and functionality.

f. Eco Tilling: Eco-Tilling (Targeting Induced Original Lesions in Genomes) is a modified tillage approach for identifying single-nucleotide allelic variants (or, more accurately, compelling point mutations) in target genes. It is a viable alternative. Mutations in the gene of interest are identified [11]. A large number of individual variations may be found quickly, and he only needs to sequence one existing variation once for each haplotype, which saves money. In tillage, chemical mutagens are employed to introduce random mutations. To induce a single-nucleotide point mutation (G/C to A/T) in seeds, EMS, a chemical mutagen, is used [12]. M1 seeds self-pollinate to produce the M2 seed population. Look for point mutations in M2 progeny from a single seed line. During screening, DNA is pooled in a variety of ways to improve the effectiveness of mutation detection. 5′ prime end-specific primers are used to target the locus of interest, and the PCR products are heated and chilled to form heteroduplexes. The CEL I nuclease enzyme is used to cleave the base mismatch, and the products reflecting the generated mutations are examined using denaturing polyacrylamide gel electrophoresis. These PCR products are denatured and reannealed to allow for mismatched or heteroduplex conformations, which reflect natural and compelling SNPs.

5. Steps involved in designing KASP markers

Designing Kompetitive Allele-Specific PCR (KASP) markers for plant breeding involves the following steps.

The initial stage in the development of KASP markers involves the identification of the target SNP (Single Nucleotide Polymorphism). This entails determining the particular SNP that is linked to the trait or characteristic under investigation or desired for selection in the field of plant breeding. The single nucleotide polymorphism (SNP) in question should possess biological significance and demonstrate relevance to the specific breeding objectives at hand [9].

5.1 SNP validation

Following the identification of the target single nucleotide polymorphism (SNP), it is imperative to undertake a process of validation to ascertain its genuine association with the specific trait of interest. The validation process may entail genotyping a varied collection of plant samples in order to verify the presence of the single nucleotide polymorphism (SNP) and its correlation with the specific trait. The process of primer design holds significant importance in the context of KASP assays, as it involves the creation of primers that are unique to the target of interest. The amplification of the area surrounding the target SNP in KASP markers is achieved through the utilization of allele-specific primers. To amplify each single nucleotide polymorphism (SNP), it is necessary to utilize a pair of allele-specific primers, with one primer designed for each allele, often representing the main and minor alleles. The design of these primers should be allele-specific, ensuring that they exclusively amplify the targeted allele [13, 14].

5.2 Fluorescent labelling

Within KASP tests, it is customary to employ distinct fluorescent dyes (e.g., FAM for one allele and HEX for the other) to label the allele-specific primers. The process of labelling facilitates the discrimination of alleles during the polymerase chain reaction (PCR) amplification. This study aims to determine the best polymerase chain reaction (PCR) settings for the KASP assay. This entails the determination of the annealing temperature, cycle parameters, and additional conditions in order to guarantee the targeted amplification of the single nucleotide polymorphism (SNP) area. Control markers are frequently used into KASP experiments. The aforementioned control markers are markers that lack polymorphism and serve to amplify a specific and well-defined section of the genome. Positive controls are utilized in order to verify the proper functioning of the polymerase chain reaction (PCR) process and to standardize the fluorescence measurements.

5.3 Testing and optimization

Prior to implementing KASP markers in high-throughput genotyping, it is imperative to do thorough testing and optimize the assay. To ensure the reliability and robustness of the outcomes, it may be necessary to conduct experiments involving the examination of various primer concentrations, annealing temperatures, and cycle circumstances.

5.4 Genotyping

Following the validation and optimization of the KASP test, it can be employed for genotyping a greater quantity of plant samples. The experimental procedure encompasses the extraction of DNA from plant tissue, followed by PCR amplification with the KASP assay, and subsequently assessing the fluorescence data to ascertain the genotype of each individual sample.

5.5 Data analysis

The genotyping data will be subjected to analysis in order to ascertain the allelic composition of each individual sample. The existence of specific alleles can be determined by analyzing the fluorescence signals emitted by different dyes.

5.6 Data interpretation

Analyze the genotyping data within the framework of the breeding objectives. The objective is to identify the genotypes that are linked to the desired trait or characteristic, and thereafter utilize this knowledge to make informed decisions in the selection of plants for further breeding purposes.

5.7 Marker validation

The aim of this study is to assess the efficacy of the KASP marker in larger breeding populations, hence ensuring its reliability in consistently selecting for the desired trait.

5.8 Incorporation into breeding programs

Following validation, the KASP marker should be integrated into the plant breeding program. Marker-assisted selection (MAS) can be employed to effectively identify plants possessing the required genotype, hence expediting the breeding procedure.

5.9 Ongoing surveillance

It is advisable to consistently observe and evaluate the efficacy of KASP markers within your breeding program, as the correlation between the single nucleotide polymorphism (SNP) and the specific characteristic under consideration may undergo alterations over time as a result of recombination or other influencing variables.

6. Bioinformatics tools used in allele mining

Allele mining is a fundamental procedure in the field of genetics and genomics, whereby the objective is to discern and elucidate the allelic variations present within a given population of organisms. This method is commonly employed to gain insights into the genetic diversity within a population or to ascertain the presence of alleles that may be linked to particular traits of significance. The field of bioinformatics plays a pivotal part in the process of allele mining by offering a wide range of tools and resources that facilitate the study and interpretation of data (**Table 1**). The following is a compilation of frequently employed bioinformatics tools utilized in the process of allele mining.

S.No.	Tool	Function	URL
1	MEME	Multiple EM for Motif Elicitation. Analyzing DNA and protein sequence motifs for similarities among them.	meme.nbcr.net/meme/cgi-bin/meme. CGI [15]
2	JASPAR	Database of Transcription Factor Binding Site (TFBS)	http://jaspar.genereg.net/ [16]
3	AGRIS	Arabidopsis gene regulatory information server. A new information resource of Arabidopsis cis- promoter	http://arabidopsis.med.ohio-state.edu [17]
4	FastPCR	FastPCR software for PCR primers or probes design and in silicoPCR, oligonucleotide assembly	http://en.bio-soft.net/pcr/FastPCR.html [18]
5	PlantCARE	A database of plant cis-acting regulatory elements	http://bioinformatics.psb.ugent.be/webtools/plantcare/html [19]
6	Primer 3	Primer design. A computer program that suggests PCR primers for a variety of applications	http://frodo.wi.mit.edu/primer3/ [20]
7	Plantprom DB	A database of plant promoter sequences	http://mendel.cs.rhul.ac.uk/mendel.php?topic=plantprom [21]
8	Clustal Omega	Multiple alignment of DNA and protein sequences/Multiple sequence alignment programs.	https://www.ebi.ac.uk/Tools/msa/clustalo/ [22]
9	T-Coffee	Multiple alignment of DNA and protein sequences/Multiple sequence alignment programs.	https://www.ebi.ac.uk/Tools/msa/tcoffee/ [23]
10	MAFFT	Multiple alignment of DNA and protein sequences/Multiple sequence alignment programs.	https://www.ebi.ac.uk/Tools/msa/mafft/ [24]
11	MUSCLE	Multiple alignment of DNA and protein sequences/Multiple sequence alignment programs.	https://www.ebi.ac.uk/Tools/msa/muscle/ [25]
12	DnaSP	DNA Sequence Polymorphism analysis, to identify SNPs and InDels and Phylogeny analysis	http://www.ub.edu/dnasp/ [26]
13	MEGA	Integrated Software Molecular Evolutionary Genetics Analysis and Sequence Alignment.	https://www.ncbi.nlm.nih.gov/pmc/articles/PMC2562624/ [27]
14	Transcriptional regulatory element database (TRED)	Collection of mammalian regulatory elements	http://rulai.cshl.edu/cgi-bin/TRED/tred.cgi?process=home [28]
15	Multalin	Multiple alignment of DNA and protein sequences/Multiple sequence alignment programs.	http://multalin.toulouse.inra.fr/multalin/ [29]
16	MAT inspector	To predict TFBS and of promoter analysis	http://www.genomatix.de/products/matinspector/ [28]

Table 1.
List and details of different bioinformatics tools used in allele mining.

7. KASP markers in plant breeding

The utilization of the Kompetitive Allele-Specific PCR (KASP) assay has significantly advanced our understanding and improvement of various traits in cereal crops. Recent studies from 2017 onwards have demonstrated the efficacy of the KASP assay in identifying key genetic markers and genes associated with important traits (**Table 2**).

In wheat, researchers have made notable discoveries using the KASP assay. Wang et al. [30] and Muellner et al. [63] identified two QTLs associated with dwarf bunt resistance, while Jiang et al. [31] discovered the QFhb-5A marker for Fusarium head blight resistance. Additionally, Li et al. [32] and Liu et al. [33] identified markers associated with leaf rust resistance and stripe rust resistance, respectively. Other studies investigated traits such as grain yield-related traits, green bug resistance, pre-harvest sprouting resistance, kernel weight, wheat curl mite resistance, heat tolerance, and drought tolerance [34–40].

The KASP assay has also been instrumental in uncovering the genetic basis of important traits in rice. In another study he identified 12 markers associated with

Crops	Traits	Number and name of QTLs/markers/genes	References
Wheat	Dwarf bunt resistance	2 QTLs	[30]
Wheat	Fusarium head blight resistance	QFhb-5A	[31]
Wheat	Leaf rust resistance	4 markers	[32]
Wheat	Stripe rust resistance	3 markers	[33]
Wheat	Grain yield-related traits	26 markers	[34]
Wheat	Green bug resistance	Gb7 gene	[35]
Wheat	Hessian fy resistance	H32 gene	[35]
Wheat	Pre-harvest sprouting resistance	10 markers	[36]
Wheat	Kernel weight	TaTAP46-5A gene	[37]
Wheat	Wheat curl mite resistance	2 markers	[38]
Wheat	Heat tolerance	sHSP26 gene	[39]
Wheat	Drought tolerance	TaSST-D1, TaSST-A1	[40]
Rice	Narrow root cone angle	12 markers	[41]
Rice	Grain yield and adaptability traits	110 markers	[42]
Rice	Variety identification and breeding guidance	48 markers	[43]
Rice	Drought and low nitrogen tolerances	8 markers	[44]
Rice	Bacterial leaf streak resistance	xa5 gene	[45]
Rice	Cold tolerance	6 markers	[46]
Rice	Grain yield and yield components	21 markers	[47]

Crops	Traits	Number and name of QTLs/markers/genes	References
Rice	Salt tolerance	25 markers	[32]
Rice	Brown plant hopper resistance	20 markers	[48]
Rice	Dirty panicle disease resistance	12 markers	[49]
Maize	Agronomic traits including yield	50 markers	[50]
Maize	—	202 markers	[51]
Maize	Maize tar spot complex resistance	qRtsc8–1 QTL	[52]
Maize	Stalk fracture angle	2 markers	[53]
Maize	Enrichment of provitamin A content	5 markers	[54]
Maize	Chilling-tolerant	20 markers	[55]
Barley	Greenbug resistance	2 markers	[56]
Barley	Greenbug resistance	3 markers	[57]
Barley	Leaf rust resistance	Rph13 gene	[58]
Oat	Stem rust resistance	Pg13 gene	[59]
Oat	Oat crown rust resistance	Pc39 gene	[60]
Sorghum	Starch content and constitution	7 markers	[61]
Pearl millet	Pollen production	2 markers	[62]

Table 2.
The following is a list of current studies that have focused on the development of KASP markers for a variety of critical traits.

narrow root cone angle [41], while indifferent study [42] they identify 110 markers associated with grain yield and adaptability traits. According to another study Tang et al. [43] identified 48 markers for variety identification and breeding guidance. Other studies focused on drought and low nitrogen tolerances, bacterial leaf streak resistance, cold tolerance, grain yield and yield components, salt tolerance, brown planthopper resistance, and dirty panicle disease resistance [32, 44–49].

In maize, the KASP assay has played a crucial role in unraveling the genetic architecture of agronomic traits. In another study [50] they identified 50 markers associated with various agronomic traits, including yield, while in different study Lu [51] discovered 202 markers providing valuable genetic information in maize research. Studies also focused on maize tar spot complex resistance, stalk fracture angle, enrichment of provitamin A content, and chilling tolerance [52–55].

Furthermore, the KASP assay has been employed in other cereal crops. Xu [56, 57] identified markers associated with greenbug resistance in barley, and Jost [58] identified the Rph13 gene for leaf rust resistance in barley. Kebede [59] and Zhao [60] utilized the KASP assay to identify genes associated with stem rust resistance and oat crown rust resistance in oat, respectively. Chen [61] identified markers associated with starch content and constitution in sorghum, while Pucher [62] identified markers associated with pollen production in pearl millet.

These studies collectively highlight the versatility and effectiveness of the KASP assay in identifying QTLs, markers, and genes associated with various traits in cereal crops. The findings have significant implications for crop improvement programs, facilitating the development of high-yielding, disease-resistant, and stress-tolerant crop varieties.

8. Conclusion

Allele mining is a potent crop development strategy that takes advantage of natural genetic variation within plant populations. This strategy is useful for identifying and exploiting advantageous genetic variants or alleles linked to desirable qualities such as higher agricultural yields, disease resistance, stress tolerance, and nutritional quality. The incorporation of modern molecular techniques, such as the Kompetitive Allele-Specific PCR (KASP) assay, has revolutionized allele mining, allowing researchers to uncover and use genetic markers and genes associated with these desirable features more quickly.

The KASP assay has proven to be a powerful tool for finding essential genetic markers and genes linked with important agronomic features in recent research spanning a variety of cereal crops, including wheat, rice, maize, barley, oat, sorghum, and pearl millet. This not only speeds up the breeding process, but it also adds to the production of crop varieties that meet rising global food need, environmental resilience, and better nutritional value.

Author details

Hemant Sharma*, Sourabh Kumar and Deepa Bhadana
Department of Genetics and Plant Breeding, Chaudhary Charan Singh University, UP, India

*Address all correspondence to: sharmahemant150893@gmail.com

References

[1] Pandey S, Singh A, Parida SK, Prasad M. Combining speed breeding with traditional and genomics-assisted breeding for crop improvement. Plant Breeding. 2022;**141**(3):301-313

[2] Lamaoui M, Jemo M, Datla R, Bekkaoui F. Heat and drought stresses in crops and approaches for their mitigation. Frontiers in Chemistry. 2018;**6**:26

[3] Raddy AM, Gambhire VB, Raddy RT. Allele mining in crop improvement. International Journal of Development Research. 2014;**4**:300-305

[4] Kumar GR, Sakthivel K, Sundaram RM, Neeraja CN, Balachandran SM, Rani NS, et al. Allele mining in crops: Prospects and potentials. Biotechnology Advances. 2010;**28**:451-461

[5] Salgotra RK, Stewart CN Jr. Functional markers for precision plant breeding. International Journal of Molecular Sciences. 2020;**21**(13):4792

[6] Li L, Sun Z, Zhang Y, Ke H, Yang J, Li Z, et al. Development and utilization of functional Kompetitive allele-specific PCR markers for key genes underpinning fiber length and strength in Gossypium hirsutum L. Frontiers in Plant Science. 2022;**13**:853827

[7] Barkley NA, Wang ML. Application of tilling and ecotilling as reverse genetic approaches to elucidate the function of genes in plants and animals. Current Genomics. 2008;**9**(4):212-226

[8] Comai L, Young K, Till BJ, Reynolds SH, Greene EA, Codomo CA, et al. Efficient discovery of DNA polymorphisms in natural populations by ecotilling. The Plant Journal. 2004;**37**(5):778-786

[9] Margulies M, Egholm M, Altman WE, Attiya S, Bader JS, Bemben LA, et al. Genome sequencing in microfabricated high-density picolitre reactors. Nature. 2005;**437**(7057):376-380

[10] Hutchison CA. DNA sequencing: Bench to bedside and beyond. Nucleic Acids Research. 2007;**35**:6227-6237

[11] Till BJ, Reynolds SH, Green EA, Codomo CA, Enns LC, Johnson JE, et al. Large-scale discovery of induced point mutations with high-throughput tilling. Genome Research. 2003;**13**:524-530

[12] Nagy J, Sulyok D, Huzsvai L. Effect of tillage on the yield of crop plants. Cereal Research Communications. 2006;**34**(1):255-258

[13] Ma KB, Yang SJ, Jo YS, Kang SS, Nam M. Development of Kompetitive allele specific PCR markers for identification of persimmon varieties using genotyping-by-sequencing. Electronic Journal of Biotechnology. 2021;**49**:72-81

[14] Kalendar R, Shustov AV, Akhmetollayev I, Kairov U. Designing allele-specific competitive-extension PCR-based assays for high-throughput genotyping and gene characterization. Frontiers in Molecular Biosciences. 2022;**9**:773956

[15] Bailey TL, Williams N, Misleh C, Li WW. MEME: Discovering and analyzing DNA and protein sequence motifs. Nucleic Acids Research. 2006;**34**:W369-W373

[16] Bryne JC, Valen E, Tang MHE, Marstrand T, Winther O, Piedade ID, et al. JASPAR, the open access database of transcription factor-binding profiles: New content and tools in the

2008 update. Nucleic Acids Research. 2008;**36**:D102-D106

[17] Davuluri RV, Sun H, Palaniswamy SK, Matthews N, Molina C, Kurtz M, et al. AGRIS: Arabidopsis gene regulatory information server, an information resource of Arabidopsis cis-regulatory elements and transcription factors. BMC Bioinformatics. 2003;**4**:25

[18] Kalendar R, Lee D, Schulman AH. FastPCR software for PCR primer and probe design and repeat search. Genes, Genomes and Genomics. 2009;**3**(1):1-4

[19] Lescot M, Dehais P, Thijs G, Marchal K, Moreau Y, YVD P, et al. PlantCARE, a database new entries and other development. Nucleic Acids Research. 2007;**35**:D137-D140

[20] Rozen S, Skaletsky H. Primer3 on the WWW for general users and for biologist programmers. Bioinformatics Methods and Protocols. 1999;**132**:365-386

[21] Shahmuradov IA, Gammerman AJ, Hancock JM, Bramley PM, Solovyev VV. PlantProm: A database of plant promoter sequences. Nucleic Acids Research. 2003;**31**(1):114-117

[22] Sievers F, Wilm A, Dineen D, Gibson TJ, Karplus K, Li W, et al. Fast, scalable generation of high-quality protein multiple sequence alignments using Clustal omega. Molecular Systems Biology. 2011;7:539

[23] Notredame C, Higgins DG, Heringa J. T-Coffee: A novel method for fast and accurate multiple sequence alignment. Journal of Molecular Biology. 2000;**302**(1):205-217

[24] Katoh K, Misawa K, Kuma K-i, Miyata T. MAFFT: A novel method for rapid multiple sequence alignment based on fast Fourier transform. Nucleic Acids Research. 2002;**30**(14):3059-3066

[25] Edgar RC. Muscle: A multiple sequence alignment method with reduced time and space complexity. BMC Bioinformatics. 2004;**5**(1):1-9

[26] Rozas J, Ferrer-Matta A, Ramos-Onsins SE, Sanchez-Gracia A. DnaSP 6: DNA sequence polymorphism analysis of large data sets. Molecular Biology and Evolution. 2017;**32**(12):3299-3302

[27] Kumar S, Nei M, Dudley J, Tamura K. MEGA: A biologist-centric software for evolutionary analysis of DNA and protein sequences. Briefings in Bioinformatics. 2008;**9**(4):299-306

[28] Jiang C, Xuan Z, Zhao F, Zhang MQ. TRED: A transcriptional regulatory element database, new entries and other development. Nucleic Acids Research. 2007;**35**(Suppl. 1):D137-D140

[29] Mitchell C. MultAlin–multiple sequence alignment. Bioinformatics. 1993;**9**(5):614

[30] Wang R, Gordon T, Hole D, Zhao W, Isham K, Bonman JM, et al. Identification and assessment of two major QTLs for dwarf bunt resistance in winter wheat line 'IDO835'. Theoretical and Applied Genetics. 2019;**132**:2755-2766

[31] Jiang P, Zhang X, Wu L, He Y, Zhuang W, Cheng X, et al. A novel QTL on chromosome 5AL of Yangmai 158 increases resistance to Fusarium head blight in wheat. Plant Pathology. 2020;**69**(2):249-258

[32] Li Z, Yuan C, Herrera-Foessel SA, Randhawa MS, Huerta-Espino J, Liu D, et al. Four consistent loci confer adult plant resistance to leaf rust in the durum wheat lines Heller# 1 and Dunkler. Phytopathology. 2020;**110**(4):892-899

[33] Liu Y, Qie Y, Li X, Wang M, Chen X. Genome-wide mapping of quantitative

trait loci conferring all-stage and high-temperature adult-plant resistance to stripe rust in spring wheat landrace PI 181410. International Journal of Molecular Sciences. 2020;**21**(2):478

[34] Yang L, Zhao D, Meng Z, Xu K, Yan J, Xia X, et al. QTL mapping for grain yield-related traits in bread wheat via SNP-based selective genotyping. Theoretical and Applied Genetics. 2020;**133**:857-872

[35] Tan CT, Yu H, Yang Y, Xu X, Chen M, Rudd JC, et al. Development and validation of KASP markers for the greenbug resistance gene Gb7 and the hessian fly resistance gene H32 in wheat. Theoretical and Applied Genetics. 2017;**130**:1867-1884

[36] Liu G, Mullan D, Zhang A, Liu H, Liu D, Yan G. Identification of KASP markers and putative genes for pre-harvest sprouting resistance in common wheat (Triticum aestivum L.). The Crop Journal. 2023;**11**(2):549-557

[37] Zhang Y, Li T, Geng Y, Wang Y, Liu Y, Li H, et al. Identification and development of a KASP functional marker of TaTAP46-5A associated with kernel weight in wheat (Triticum aestivum). Plant Breeding. 2021;**140**(4):585-594

[38] Dhakal S, Tan CT, Anderson V, Yu H, Fuentealba MP, Rudd JC, et al. Mapping and KASP marker development for wheat curl mite resistance in "TAM 112" wheat using linkage and association analysis. Molecular Breeding. 2018;**38**:1-3

[39] Comastri A, Janni M, Simmonds J, Uauy C, Pignone D, Nguyen HT, et al. Heat in wheat: Exploit reverse genetic techniques to discover new alleles within the Triticum durum sHsp26 family. Frontiers in Plant Science. 2018;**9**:1337

[40] Khalid M, Afzal F, Gul A, Amir R, Subhani A, Ahmed Z, et al. Molecular characterization of 87 functional genes in wheat diversity panel and their association with phenotypes under well-watered and water-limited conditions. Frontiers in Plant Science. 2019;**10**:717

[41] Vinarao R, Proud C, Snell P, Fukai S, Mitchell J. QTL validation and development of SNP-based high throughput molecular markers targeting a genomic region conferring narrow root cone angle in aerobic rice production systems. Plants. 2021;**10**(10):2099

[42] Sandhu N, Singh J, Singh G, Sethi M, Singh MP, Pruthi G, et al. Development and validation of a novel core set of KASP markers for the traits improving grain yield and adaptability of rice under direct-seeded cultivation conditions. Genomics. 2022;**114**(2):110269

[43] Tang W, Lin J, Wang Y, An H, Chen H, Pan G, et al. Selection and validation of 48 KASP markers for variety identification and breeding guidance in conventional and hybrid rice (Oryza sativa L.). Rice. 2022;**15**(1):48

[44] Feng B, Chen K, Cui Y, Wu Z, Zheng T, Zhu Y, et al. Genetic dissection and simultaneous improvement of drought and low nitrogen tolerances by designed QTL pyramiding in rice. Frontiers in Plant Science. 2018;**9**:306

[45] Thianthavon T, Aesomnuk W, Pitaloka MK, Sattayachiti W, Sonsom Y, Nubankoh P, et al. Identification and validation of a qtl for bacterial leaf streak resistance in rice (Oryza sativa l.) against thai xoc strains. Genes. 2021;**12**(10):1587

[46] Yang L, Liu H, Lei L, Wang J, Zheng H, Xin W, et al. Combined QTL-sequencing, linkage mapping, and RNA-sequencing identify candidate genes and KASP markers for low-temperature germination in Oryza sativa L. ssp. Japonica. Planta. 2023;**257**(6):1-3

[47] Ashfaq H, Rani R, Perveen N, Babar AD, Maqsood U, Asif M, et al. KASP mapping of QTLs for yield components using a RIL population in basmati rice (Oryza sativa L.). Euphytica. 2023;**219**(7):79

[48] Ishwarya Lakshmi VG, Sreedhar M, JhansiLakshmi V, Gireesh C, Rathod S, Bohar R, et al. Development and validation of diagnostic KASP markers for brown planthopper resistance in Rice. Frontiers in Genetics. 2022;**13**:914131

[49] Riangwong K, Aesomnuk W, Sonsom Y, Siangliw M, Unartngam J, Toojinda T, et al. QTL-seq identifies genomic regions associated with resistance to dirty panicle disease in rice. Agronomy. 2023;**13**(7):1905

[50] Chen Z, Tang D, Ni J, Li P, Wang L, Zhou J, et al. Development of genic KASP SNP markers from RNA-Seq data for map-based cloning and marker-assisted selection in maize. BMC Plant Biology. 2021;**21**:1-1

[51] Lu H, Zhou L, Lin F, Wang R, Wang F, Zhao H. Development of efficient KASP molecular markers based on high throughput sequencing in maize. Acta Agronomica Sinica. 2019;**45**(6):872-878

[52] Ren J, Wu P, Huestis GM, Zhang A, Qu J, Liu Y, et al. Identification and fine mapping of a major QTL (qRtsc8-1) conferring resistance to maize tar spot complex and validation of production markers in breeding lines. Theoretical and Applied Genetics. 2022;**135**(5):1551-1563

[53] Wang X, Shi Z, Zhang R, Sun X, Wang J, Wang S, et al. Stalk architecture, cell wall composition, and QTL underlying high stalk flexibility for improved lodging resistance in maize. BMC Plant Biology. 2020;**20**(1):1-2

[54] Kebede D, Mengesha W, Menkir A, Abe A, Garcia-Oliveira AL, Gedil M. Marker based enrichment of provitamin a content in two tropical maize synthetics. Scientific Reports. 2021;**11**(1):14998

[55] Yan M, Li F, Sun Q, Zhao J, Ma Y. Identification of chilling-tolerant genes in maize via bulked segregant analysis sequencing. Environmental and Experimental Botany. 2023;**208**:105234

[56] Xu X, Mornhinweg D, Bai G, Li G, Bian R, Bernardo A, et al. Characterization of Rsg3, a novel greenbug resistance gene from the Chinese barley landrace PI 565676. The Plant Genome. 2023;**16**(1):e20287

[57] Xu X, Mornhinweg D, Bernardo A, Li G, Bian R, Steffenson BJ, et al. Characterization of Rsg2. a3: A new greenbug resistance allele at the Rsg2 locus from wild barley (Hordeum vulgare ssp. spontaneum). The Crop Journal. 2022;**10**(6):1727-1732

[58] Jost M, Singh D, Lagudah E, Park RF, Dracatos P. Fine mapping of leaf rust resistance gene Rph13 from wild barley. Theoretical and Applied Genetics. 2020;**133**:1887-1895

[59] Kebede AZ, Admassu-Yimer B, Bekele WA, Gordon T, Bonman JM, Babiker E, et al. Mapping of the stem rust resistance gene Pg13 in cultivated oat. Theoretical and Applied Genetics. 2020;**133**:259-270

[60] Zhao J, Kebede AZ, Bekele WA, Menzies JG, Chong J, Mitchell Fetch JW, et al. Mapping of the oat crown rust resistance gene Pc39 relative to single nucleotide polymorphism markers. Plant Disease. 2020;**104**(5):1507-1513

[61] Chen BR, Wang CY, Ping WA, Zhu ZX, Ning XU, Shi GS, et al.

Genome-wide association study for starch content and constitution in sorghum (Sorghum bicolor (L.) Moench). Journal of Integrative Agriculture. 2019;**18**(11):2446-2456

[62] Pucher A, Hash CT, Wallace JG, Han S, Leiser WL, Haussmann BI. Mapping a male-fertility restoration locus for the a 4 cytoplasmic-genic male-sterility system in pearl millet using a genotyping-by-sequencing-based linkage map. BMC Plant Biology. 2018;**18**:1-1

[63] Muellner AE, Eshonkulov B, Hagenguth J, Pachler B, Michel S, Buerstmayr M, et al. Genetic mapping of the common and dwarf bunt resistance gene Bt12 descending from the wheat landrace PI119333. Euphytica. 2020;**216**:1-5

Chapter 3

Molecular Selection Tools in Adaptive Phenology of *Populus trichocarpa* Breeds for the Nordic-Baltic Region

Anneli Adler, Almir Karacic, Rami-Petteri Apuli, Ann-Christin Rönnberg Wästljung, Magnus Hertzberg, Martin Weih and Pär K. Ingvarsson

Abstract

Fast-growing poplars have the potential to improve the biomass supply required for the transition to bio-based economies in the Nordic-Baltic region. As early successional trees, poplars are efficient biomass producers in relatively short rotations, when high-yielding, climate-adapted clones are available for commercial deployment. In Sweden, poplar breeding focused on adapting *Populus trichocarpa* to the Swedish climate by crossing parents from distant populations along latitudinal and maritime-continental clines on the Pacific coast of North America. Clonal trials with progeny from these crosses were established in the Nordic-Baltic region. Elite individuals in terms of stemwood production were used to identify candidate genes for adaptation to local photoperiod and climate in the region. The next breeding cycle utilized the elite individuals in the clonal trials to generate a training population. Genomic selection of the progeny in the training population will facilitate early selection of poplar clones for commercial deployment in the Nordic-Baltic region and reduce the time required for successive plant breeding cycles.

Keywords: adaptive phenology, alleles, biomass, bud burst, bud set, candidate genes, forest industry, genomic prediction model, genomic selection, heritability, hybrid poplar, genetic marker density, nucleotides, *Populus trichocarpa*, stem wood, training population, woody biomass, yield

1. Introduction

When breeding poplars for northern latitudes, significant attention is given to the adaptation of their phenological traits to regional climates [1–4]. The most important phenological events that distinguish the period of dormancy and active growth are bud flushing, growth cessation, and bud set [5]. The timing of these events is crucial for an individual genotype to adjust its growth efficiently to the temperature and photoperiod in a given region [6]. Thus, different populations within the species' geographic range are set under intense selection pressure to adapt to latitudinal and altitudinal gradients of daylight and temperature. Individuals with phenologies that do not match their environments are at a higher risk of being frost-injured, especially during the transition periods from autumn to winter and winter to spring. However, prolonged growth can also lead to a competitive advantage by allowing for greater utilization of the growing season [7, 8]. As a result, the phenology of each population is shaped by its adaptation to the local and regional environments but still contains significant genetic variation for phenological traits [9].

Adaptation of a species to new geographic regions relies on the variation within and between populations. *Populus trichocarpa* (Torr & A. Gray ex Hook) offers ample opportunities for adaptation to the Nordic-Baltic region due to its wide ecological range and geographic distribution [10]. While phenological traits in natural populations adjust to climate extremes, the artificial transition is typically directed northward to extend the active growth period and increase biomass production [11]. In species like Norway spruce, population-level variation mitigates the risks of this transition. In operative poplar cultivation, however, individual clones lack the buffer contained in the variation within a deployed population, which increases risks for financial losses.

The Nordic-Baltic forest industry and energy sector have used native aspen (*P. tremula* L.) for decades, and it is reasonable to assume that poplars can also be utilized in current and innovative manufacturing processes. Cultivated poplars are a marginal resource in the region, and widespread cultivation would require an extensive breeding and testing program. However, molecular selection tools offer a faster and more cost-efficient way to select clones, which is particularly useful for poplars, where clone selection and testing costs are disproportionally high, considering their marginal role as a raw-material asset in the region.

For over 30 years, the Swedish University of Agricultural Sciences has been running a climate adaptation program for *P. trichocarpa*. Since 2007, various stakeholders have also been included in research and testing activities. The most promising clones have been planted in pilot plantations throughout Sweden, Lithuania, Latvia, and Estonia, and several clones are expected to be registered with the Swedish Board of Forestry by 2024.

2. Potential of hybrid poplars in the Nordic-Baltic wood market

Forests offer multiple ecosystem services and are a crucial resource in the transition to biobased economies. This is especially true for the Nordic-Baltic region, which has abundant forest resources and well-established wood industries. Consequently, the region is a major hub for education, research and development, testing, and production of innovative wood-based products, including textiles, biofuels, biochemicals, and engineered wood products for the construction sector [12]. However, the increasing demand for wood as a raw material is likely to be in conflict with the environmental role of forest ecosystems as carbon sinks and guardians of biodiversity [13]. Additionally, in a warming climate, forests may be more susceptible to severe droughts, fires, storms, diseases, and insect outbreaks, which could further reduce the availability of woody biomass [14].

Expanding the area of short-rotation poplar plantations is a viable option for increasing the stocks of industrial wood. Poplar plantations are usually designed for high biomass production enabling relatively frequent financial returns for the growers. This type of plantations increase the value of low-quality and marginal arable land, making them attractive investment targets for capital investors. Many EU countries have reported their anticipation of heightened private investments in fast-growing tree species [15].

Populus species worldwide cover an estimated 60 million hectares, with a significant portion located in Canada and the Russian Federation in the form of natural aspen forests [15]. Planted poplars account for 8.3 million hectares, grown primarily for the production of industrial roundwood (5 million ha), environmental protection (1.9 million ha), and fuelwood (0.8 million ha). The main products derived from these plantations include veneer, plywood, pulpwood, wood panels, furniture, and matches. In Europe, poplar plantations occupy approximately 0.8 million ha, with Italy harvesting 50% of its annual hardwood biomass from poplar plantations.

In the Nordic-Baltic region, the current area of poplar plantations is approximately 5000 ha. Hybrid aspen is planted on an additional 12,000 ha. Consequently, *Populus* plantations are a minor source of raw material. Native aspen (*Populus tremula* L.), on the other hand, represents a valuable wood resource for pulping, match production, furniture, particle boards, packaging, biofuels, and many other wood-based products. Estonian Cell, for example, has a modern facility built in 2006 for processing Baltic aspen into bleached chemi-termo-mechanical pulp. They produce 170, 000 tons of pulp annually, requiring 450,000 m^3 of aspen pulpwood [16]. This corresponds to a harvest of about 1000 ha and a total area of 25,000 ha of hybrid aspen plantations grown in a 25-year rotation. The Swedish Södra Cell pulp factory also uses aspen in a mix with birch and beech to produce dissolving pulp for textiles [17]. Pulpwood from Swedish poplar plantations is also used in Södra Cells's raw-material mix.

Among the Baltic countries, Lithuania has fewer forest resources than Estonia and Latvia but more set aside arable land suitable for conversion to poplar plantations. Lithuania and Latvia are now major producers of wood-based panels in the region and supply global companies such as IKEA (**Figure 1**). Estonia and Latvia also produce large quantities of pellets, mainly for export to Denmark, Great Britain, and the USA (**Figure 2**) [19]. Poplar wood is ideal for making products like oriented strand boards

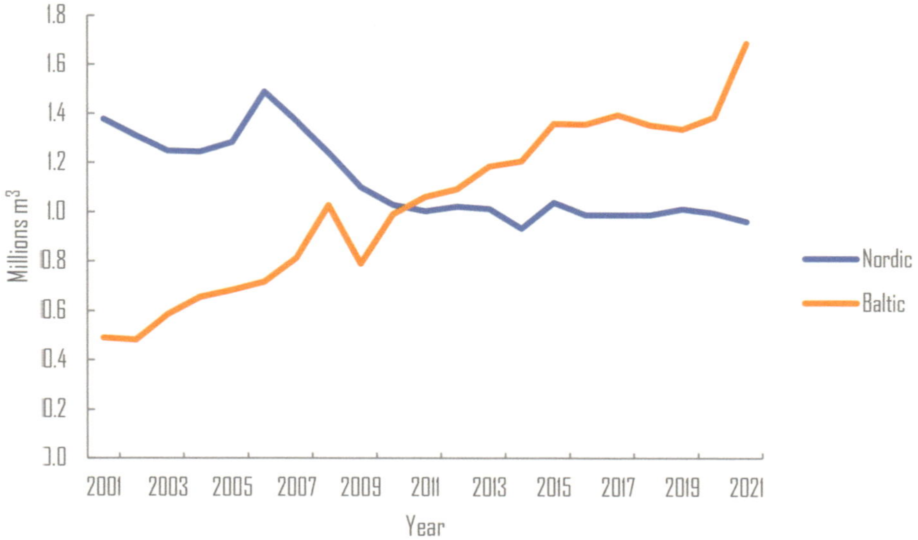

Figure 1.
Particle board production in the Nordic-Baltic countries since 2001. Nordic countries—Denmark, Finland, and Sweden. Baltic countries—Estonia, Latvia, and Lithuania [18].

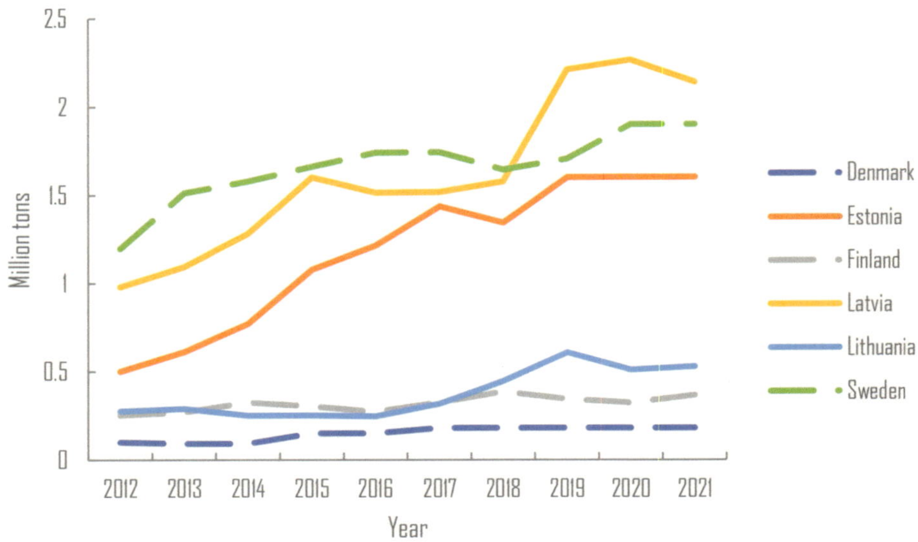

Figure 2.
Production of wood pellets in six Nordic-Baltic countries for the period 2012–2021 [18].

(OSB) and medium-density fiberboards (MDF). These products represent a significant market opportunity for poplar wood in the region. While poplars are not likely to be used for veneer in the Baltic countries, they can be utilized for inner layers in plywood, while birch veneer is used for the surface layers.

In the Nordic-Baltic countries, poplar plantations were initially established as a part of land conversion programs and for bioenergy purposes. Sweden has seen an increase in the area of hybrid poplar and hybrid aspen plantations in the past few decades, reaching approximately 4000 ha [20]. At the same time, the area of willow coppice has decreased from 14,000 ha in 2001 to just over 7000 h in 2016 [21]. One of the main reasons for the decrease in willow coppice is the significant increase in cereal prices since 2007. Similarly, the decline in Italian poplar cultivation has also been related to the profitability of alternative land use [22]. The potential of poplars as a raw-material resource in the region depends directly on the expansion of plantations on arable or other non-forested land. Fertile agricultural land is more likely to be utilized for cultivation of food and feed crops, while arable land and grasslands of poor quality, or fields that are difficult to access with agricultural machinery can be available for poplar plantations. However, grasslands can be a valuable form of land use, particularly on organic soils, providing high biodiversity and significant carbon storage potential [23]. Poplar plantations can also expand by replacing forest tree species planted on arable land during the early conversions in the mid-twentieth century. This replacement is more of a resilience issue in a changing climate rather than an opportunity to increase biomass production significantly. Despite these limitations, it has been suggested that between 1.8 and 4.6 million ha of land in the Nordic-Baltic region could be available for conversion to poplar plantations [24, 25].

3. Biology and ecology of *Populus*

The *Populus* genus comprises poplars and aspens, which can be found in various climatic conditions in the northern hemisphere, ranging from the sub-tropical regions to the Arctic [10]. Typically dioecious, wind-pollinated trees, poplars start to flower when 6–7 years old, producing large amounts of tiny, wind-spread cottony seeds. Poplars and aspens are pioneer species invading the fresh mineral deposits in riparian environments (poplars including white poplars from section *Populus,* see below) or on fire-disturbed forest sites (aspens). They also reproduce clonally through detached branches, stump shoots, and root suckers, which allows them to persist on already occupied sites. Eventually, secondary successional species will take over the sites pioneered by poplars and aspens. The exception is frequently disturbed riparian sites, where poplars and willows may regenerate cyclically. On the other hand, aspens may form pure or mixed forest stands but usually grow in small groves at forest edges.

The taxonomy of *Populus* is rather complex, the complexity being a result of low barriers to gene flow between species and even between sections of the genus leading to the emergence of intermediate forms (hybrid swarms) among sympatric

Figure 3.
Natural distribution of Populus trichocarpa (Source: [27]).

poplar species [26]. Out of the six sections in the *Populus* genus, only three—*Aigeiros*, *Tacamahaca,* and *Populus* (previously known as *Leuce)*—contain commercially valuable species. Interspecific hybridization within sections is common but is rare or absent between sections, except for *Aigeiros* (black poplars) and *Tacamahaca* (balsam poplars), which can easily be crossed, producing valuable hybrids.

Black cottonwood (*Populus trichocarpa*) is a commercially essential species in the *Tacamahaca* section. It is naturally found in the western regions of North America, from California in the south to Alaska in the north (**Figure 3**). Black cottonwood typically grows in moist environments on moraines, alluvial soils, and riparian habitats. It can also thrive in interior valleys and canyons up to 2000 m in altitude. Strait stems, narrow crowns, and impressive dimensions, reaching heights over 40 m and more than 1 m in diameter [28], characterize the species. The species is easy to propagate with dormant stem cuttings. Harvested trees can resprout from stumps or even produce root suckers.

4. Nordic-Baltic poplar cultivation practices

Operative poplar cultivation in the Nordic-Baltic region was first introduced during the early 1990s. Typically, plantations were established at a 3 × 3 m spacing ($\approx$ 1000 trees ha^{-1}), aiming at a rotation of 20 years (**Figure 4**). In forest-dominated areas of central Sweden, planting densities ranging from 1100 to 1700 trees per hectare are also used, with higher densities serving as a buffer against losses due to the browsing pressure. The great majority of Swedish poplar plantations were established with clone 'OP42' (*P. maximowiczii* × *trichocarpa*) bred in the USA at the beginning of the 1930s. In Lithuania, approximately 1700 ha of poplar plantations have been planted since 2014, mostly using a 3 × 2 m spacing (1600 trees per hectare), with the expected rotation period of no longer than 10 years (**Figure 5**).

The intensity of work to prepare land for planting varies depending on its previous use but typically involves using herbicides and cultivating the soil. Various types of planting materials are used, ranging from rooted plants (either bare root or containerized) to cuttings of varying lengths, including unrooted 2-year-old poles. In some cases, cuttings are planted through degradable plastic mulch that can be replaced by biodegradable mulch paper. However, most Nordic and Baltic entrepreneurs rely on using robust planting material and mechanical weed control after planting.

Poplar growers in the Nordic-Baltic region face specific challenges due to the relatively short growing season and cold winters. Poplar clones with better frost hardiness are required compared to the material available from central/southern European breeding programs [3]. Moreover, the market for poplar wood is limited, which makes it difficult to justify cost-intensive tending. To overcome these obstacles, poplar growers need to optimize management by combining denser initial spacing, more robust plant material, and less frequent mechanical weeding. Additional challenges are present due to high browsing and rodent populations, particularly in small, isolated poplar plantations in forest-dominated landscapes.

Site and clone selection are the two most important decisions in poplar silviculture. Poplars perform best on deep, moist, well-drained, and light-textured soils. Dry sandy soils and waterlogged soils are unsuitable as they can result in poor growth and establishment. However, the productivity of dry soils can be improved by fertilization and irrigation with wastewater, while waterlogged soils can be ditched. Wet sites are

Figure 4.
A 19-year-old trial with climate-adapted black cottonwood (P. trichocarpa) grown in Central Sweden at 59°N. The trial contains more than 100 clones, with the three best clones reaching a mean diameter of 400–450 mm at breast height. The chapter's main author for scale beside a tree of clone 722.16. The trial was established with 3.5 × 3.5 m spacing and thinned systematically after 9 years. Several candidate clones for commercial deployment are currently planted in pilot plantations and operational plantations on agricultural soils in Sweden, Latvia, Estonia, Lithuania, and Romania. The best clones will be registered at the Swedish Forest Agency in 2024. Photo by Almir Karačić.

Figure 5.
Seven-year-old poplar plantation in Lithuania established at 3 × 2 m spacing. The clone used was the Italian AF7, which suffered from repeated frost injuries on this site. This stand was established using 1.5 m long poles and harvested a couple of months after the photo was taken. Lithuanian poplar plantations are established without the application of herbicides and have been certified with FSC standards since August 2023 [29]. Photo by A. Karačić.

typically located in lowland terrain on organic soils with abundant ground vegetation. There is a relative abundance of these sites in the Nordic-Baltic region, so farmers prioritize these areas when converting land into tree plantations. However, there is a high likelihood of frequent frost, high browsing pressure, vole populations, and serious competition from herbaceous vegetation occurring on these sites. Combined, these factors represent a severe challenge to poplar growers who must employ simultaneous measures to be successful, including heavy soil preparation (mounding, for example), herbicide treatments, frequent mechanical weeding, and sometimes plant protection.

Deploying highly productive clones is the most efficient measure to increase the yields of poplar plantations. However, even for a small set of productive clones, there can be substantial variations in clonal performance [8, 30]. Some clones, such as 'OP42', are generalists and can grow well on many different soils. 'OP42' is outperformed by better-adapted clones on some sites or towards the edges of its deployment range. Planting clone mixtures is sometimes recommended to resolve the uncertainties related to clonal performance in different environments. This approach has both advantages and disadvantages, and the optimization of mixtures regarding the number and growth pattern of the clones in a mixture is necessary [31].

The Nordic-Baltic poplar plantations established with 1000 trees ha^{-1} yield total biomass at an average of 10 tons ha^{-1} $year^{-1}$ dry weight over a 20-year rotation. This corresponds to 8 tons ha^{-1} $year^{-1}$ of stem biomass or 20–25 m^3 ha^{-1} $year^{-1}$ [25]. During the second half of a rotation, individual trees face intense competition, which can impact their vitality, especially during periods of summer drought [32]. Therefore, thinning is commonly utilized to promote diameter growth and vitality in the remaining trees. The recommended final stocking of Nordic-Baltic poplar stands is 650–800 trees per hectare. The timing of thinning will depend on the initial density and site productivity, usually occurring between ages 8–14 when the mean height is 15–18 m. However, thinning poplar stands pose a risk as heavy machines can damage their shallow root system. The final harvest should be performed between December and April to ensure successful resprouting from stumps and roots. Harvesting during May–September may result in the failure of the coppice regeneration [33].

5. Current genetic resources and past breeding activities

Populus breeding and selection began in Sweden during the 1930s [34], with a focus on *P. tremula* and its hybrids with *P. tremuloides*. Initial poplar trials also included *P.* ×*euramericana* hybrids to demonstrate the general production potential of poplars in southern Sweden [35]. These activities were eventually abandoned in the 1960s. Still, a portion of the tested material was transferred into several clone archives and later used in the breeding of *P. trichocarpa* at the Swedish University of Agricultural Sciences.

In the 1970s, interest in poplars and other fast-growing species increased due to energy crises. Several trials were established in southern Sweden, mainly with material imported via Germany and Holland. Karlsson et al. [36] evaluated some of these trials, suggesting that hybrids with species from the *Tacamahaca* section were best suited for southern Swedish conditions. The Swedish Forestry Research Institute (SkogForsk) continued the work with imported poplar material, expanding testing to include agricultural fields and forest soils between 56°N and 65°N [30, 37, 38]. So far, SkogForsk has registered five poplar clones with the Swedish Forest Agency, including 'OP42'.

In the late 1980s, Ilstedt [3] conducted a test on a collection of Belgian *P.* ×*generosa* and *P. trichocarpa* in central Sweden. The material showed good growth potential but was sensitive to early autumn frost due to prolonged growth, leading to disruption of bud formation and winter hardening. The same material was tested in the milder climate of southernmost Sweden (55°40'N), where *P. ×generosa* clones failed due to stem canker [39]. Some Belgian *P. trichocarpa* clones have shown satisfactory growth compared to 'OP42' in a trial established by the Swedish west coast [8]. *P. trichocarpa* has been used in poplar breeding programs since the 1920s, producing commercially valuable hybrids in crossings with *P. maximowiczii* and/or *P. deltoides*. The hybrids with *P. deltoides* (*P. ×generosa*) have been widely employed in resource-intensive poplar cultivation in the Pacific Northwest [40]. In Europe, black cottonwood was also utilized in the Belgian poplar breeding program, producing high-yielding, leaf rust-resistant *P. ×generosa* hybrids. However, towards the end of the 1980s, the leaf rust resistance eventually collapsed, making this material unsuitable for large-scale operations. As a result, the breeding strategy was adjusted to entail tolerance rather than resistance to *Melampsora* leaf rust [41].

Recognizing the potential of *P. trichocarpa*, a project was initiated to adapt this species to the climate of central Sweden [42]. A series of more than 100 crossings involved 20 female and 30 male parent trees originating from the Pacific coastal areas of North America and more continental locations between 44°N and 60°N. Approximately 900 selected clones underwent testing for several years before planting 100 individuals in the long-term clone trial near Uppsala at 59°N. Since 2007, selected parts of this material have been tested in multiple locations in Sweden, Estonia, Latvia, Lithuania, and Poland [6, 7, 43, 44]. Since 2013, 10 candidate clones for commercial deployment have been tested in pilot plantations.

6. Adaptive phenology is one of the breeding goals

Flowering and bud flushing of poplar trees co-occur early in spring and are regulated mainly by temperature (**Figure 6**) [45]. In contrast, the timing of growth cessation is primarily governed by changes in day length and is known to have a significant genetic component (**Figure 7**) [46]. Bud formation is initiated after growth cessation and is also affected by the temperature [5]. Due to a rapid temperature cline along increasing latitudes in the Nordic-Baltic region, one of the primary goals of poplar climate adaptation is to identify genotypes with optimal timing of active growth initiation and termination.

These traits are controlled by many genes with minor effects, e.g., they are quantitatively inherited. New progeny genotypes obtained by cross-pollination of individuals from distant heterotic groups will display a spectrum of phenological characteristics. Consequently, breeding for climate adaptation involves such crosses to achieve hybrid vigor, which is a common strategy in European poplar breeding programs [47, 48]. In the Nordic-Baltic region, an example of a specific breeding goal for autumn phenology is to select clones that exhibit active growth throughout August but also have a rapid bud formation and maturation in September. This allows the new poplar clones to take advantage of favorable growing conditions with optimum temperature and rainfall in August while ensuring frost hardiness. A good benchmark for the timing of bud set in poplar clones can be developed from the Swedish horticultural growing zones (**Figure 8**), where bud formation should be accomplished approximately 1 week earlier within each successive growing zone. However, it is important to keep in mind that the timing of frost occurrence within each growing zone can vary significantly, so local conditions and premises for clone deployment should be considered.

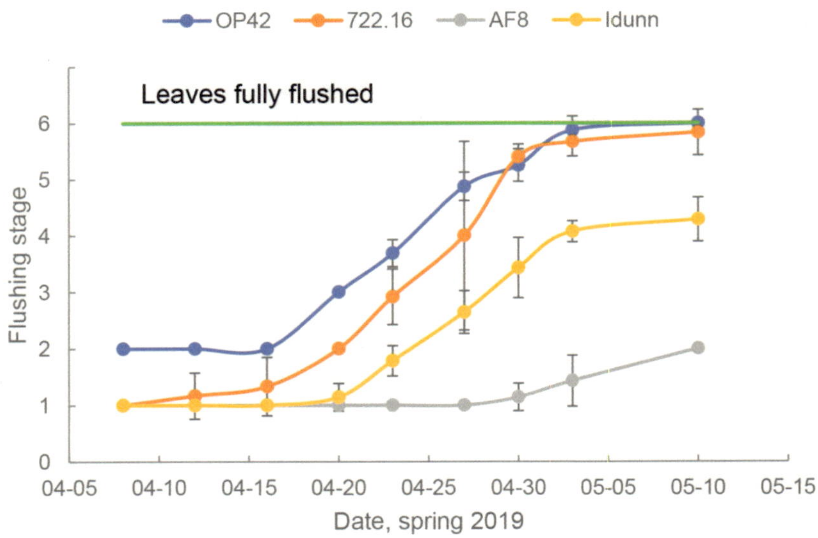

Figure 6.
Stages of bud burst (1—buds swelled but no green leaves visible; 6—leaves fully flushed with an initial shoot increment >10 mm) for four clones grown in a common garden in Uppsala, central Sweden (59°N). Clone 'OP42' (P. maximowiczii × trichocarpa) initiates growth earlier than the other three clones. Clone '722.16' is adapted to growth in central Sweden (see **Figure 4***). The Italian clone 'AF8' is not adapted to the Nordic climate and requires a higher temperature threshold to initiate bud burst. The Icelandic clone 'Idunn' is adapted to maritime conditions and will also have somewhat delayed bud flushing. The trial included 40 clones planted in 10 single-tree plots (blocks), and the number of observations for the four presented clones was 'OP42' = 4, '722.16' = 6, 'AF8' = 7, and 'Idunn' = 7. Error bars are standard deviations.*

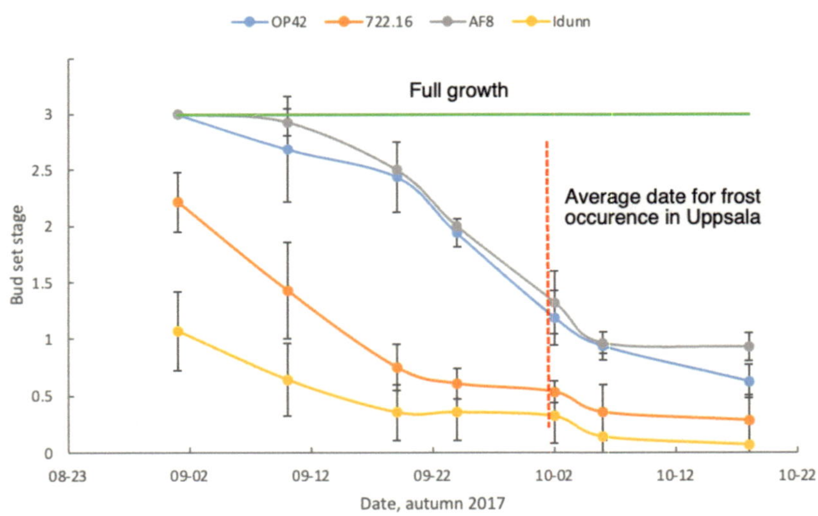

Figure 7.
Stages of bud set (3—fully growing shoots; 1.5—initiation of bud formation; 0.5—bud formed but not matured; 0—matured bud) for four clones grown in a common garden in Uppsala, Central Sweden (59°N). The growth cessation and bud set were inventoried in the year of planting. The first-year growth can be prolonged compared to the already established plants. Data from the second growing season (2018) was not useful due to an extremely dry summer causing atypic bud set in many clones, and the inventory was not possible in 2019 due to height of trees (up to 8 m). For example, the Icelandic clone 'Idunn' would normally initiate a bud set already in July. Clones 'OP42' and 'AF8' are late to set bud and will be at risk of frost damage at this latitude. Early frosts can induce frost damage to the top shoot. Cold temperatures also inhibit bud formation leading to incomplete dormancy and the risk of frost damage during the winter. The trial included 40 clones planted in 10 single-tree plots (blocks), and the number of observations for the four presented clones was 'OP42' = 4, '722.16' = 6, 'AF8' = 7, and 'Idunn' = 7. Error bars are standard deviations.

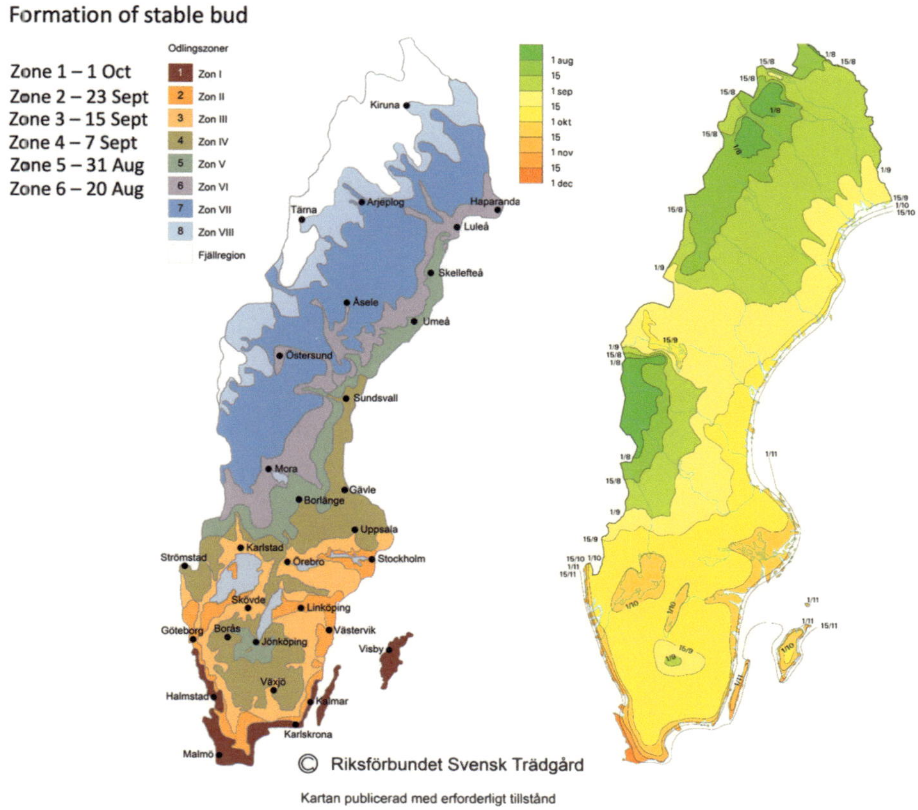

Figure 8.
The latest date for forming a stable bud is autumn within the six growing zones in Sweden. The horticultural growing zones (left) are used as guidance for clone deployment based on the timing of bud set (map from the Swedish Garden Association, with permission). The range of average dates for frost occurrence (right) is an essential benchmark for evaluating clone hardiness (map from the SMHI [49]).

Nonetheless, having a reasonable estimate of bud set for a particular clone and zone is essential, as the timing for a clone is likely to differ across different growing zones.

7. Genetic architecture of photoperiodic traits

The genetic architecture of photoperiodic traits is complex, involving many genes with minor effects. The alleles either decrease or increase the additive effect of the different genes. Many loci in the poplar genome have been identified through QTL mapping that can explain variation in bud set traits [50–59]. The identified genes regulate the active growth and dormancy cycle based on the perception of photoperiodic and dormancy signals [60]. The circadian rhythm regulates the active growth cycle, a biological process that displays an endogenous adjustable oscillation of approximately 24 h. External cues such as daylight and temperature determine the circadian rhythm in a specific environment. The interplay between various photoreceptors and the circadian clock is important for induced growth cessation and bud set in deciduous trees [61, 62]. The large number of candidate genes involved in adaptation to photoperiod and climate highlights the quantitative nature of these traits (**Figure 9**).

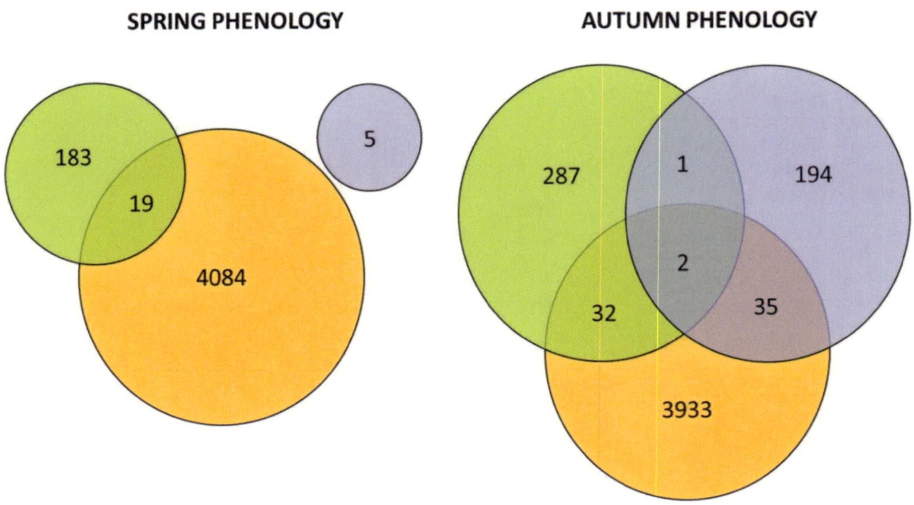

Figure 9.
Overlap between candidate genes of spring and autumn phenology in populations of Populus trichocarpa. Orange: wild accessions from Northwest coast of North America [63]. Lilac: wild accessions from western North America [64]. Green: hybrids between P. trichocarpa provenances originating from the populations along the latitudinal and longitudinal clines on the Pacific Coast of North America, bred at Swedish University of Agricultural Sciences and introduced to progeny trials in the Baltic Sea region in northern Europa [65]. Modified figure from Apuli et al. [65].

7.1 Circadian clock

Circadian clock genes are genes whose protein products generate and regulate circadian rhythms in different organisms. In poplar, several circadian clock genes have been identified: *LHY1, LHY2, TOC1, PRR5, FLOWERING LOCUS T (FT), PhyA,* etc. [52, 57]. The circadian clock resets at dawn and dusk. It involves the action of the red/far-red and blue light-receiving phytochromes (phy) and cryptochromes (cry), as well as blue light receptors from the ZEITLUPE (ZTL) family [66]. *P. trichocarpa* has one *PHYA* and two *PHYB* genes [67] which are involved in the control of growth in *Populus* [53, 68]. Many other genes are involved in the interplay of the circadian clock with the tree's growth [60].

7.2 Heritability of critical daylength

Typically, a day shorter than the specific length required for growth evokes growth cessation and bud set. The critical day length is when about 50% of individuals of a particular clone are setting buds [60]. Fabbrini et al. [46] and Rohde et al. [5] have demonstrated that traits related to the timing of growth cessation in black poplars (*P. nigra*) had significantly higher broad sense heritability (H^2 = 0.64–0.84) compared to the traits associated with the duration of bud maturation (H^2 = 0.10–0.67). Similar results have been obtained in European aspen (*P. tremula*) for populations sampled across a latitudinal gradient in Sweden [69]. The broad sense heritability for two different wild populations of *P. balsamifera* was 0.37 and 0.55 for bud flush and 0.44 and 0.72 for bud set when grown in two common gardens in Canada [45]. Several studies have used genome-wide association studies to dissect the genetic basis of growth cessation in response to critical day length in several *Populus* species [63–65, 70, 71]. Many of these studies have identified an association between the timing of growth cessation and core genes from the photoperiod pathway and the circadian clock. In particular, the gene *FLOWERING LOCUS T2* (*FT2*) has been implicated as a major effect locus in both *P. trichocarpa* [63] and *P. tremula* [70], indicating that parallel adaptation to similar environmental gradients target similar developmental pathways and sometimes even the same gene in different species.

7.3 Molecular markers for selection of operational varieties

Populus was early established as a model genus for studying the biology of woody perennials [72, 73]. This resulted in the early development of large-scale genomics resources and, ultimately, the first *de novo* draft genome for a forest tree [74]. The availability of large-scale genomics resources also resulted in early efforts aimed at studying genetic variation at the nucleotide level, first achieved by targeting short genomic regions [75–78] and ultimately through whole-genome re-sequencing [63, 79].

Thus, abundant genomic information is available for many species in the genus *Populus* that can be used for developing molecular breeding tools. The most abundant genetic marker type is single nucleotide polymorphisms (SNPs), which can be detected in the millions in most *Populus* species, e.g., Wang et al. [79]. To date, large SNP data sets are available for most species in the genus *Populus*. Methods for rapid and cost-efficient genotyping have also been developed in a few other species [80–83]. In addition, many genetic markers have also been associated with traits of interest through genome-wide association studies [64, 70, 84], making it possible to directly target variation underlying traits that are of interest as targets in e.g., breeding programs [65].

8. Timeline for the implementation of molecular markers in selection

Genomic selection (GS), initially proposed by Meuwissen et al. [85], is a method that uses genetic markers with genome-wide coverage and phenotypic information from a training population to develop a model that can be used to predict the breeding values for individuals from a breeding population with only genotype information available. GS has many advantages that increase the method's utility in modern plant breeding. First, GS can be used to perform early selection, often even at the seedling stage, without the need for extensive field testing and screenings. This facilitates early selection, which reduces the time required for successive plant breeding cycles and thus contributes to an increased genetic gain per unit of time. Second, the estimated marker effects from GS are precise, and unbiased prior marker selection is avoided [86, 87]. GS is particularly suitable for trees due to their long generation times and traits characterized by a genetic architecture consisting of many underlying genes, each with relatively small effects. Furthermore, traits of interest for tree breeding are often expensive to phenotype or are displayed late in the life cycle, further increasing the utility of GS.

Implementing a genomic selection program for tree breeding encompasses two stages; the first stage relies on a "training population" consisting of individuals that are both phenotyped and genotyped. The combined genotype and phenotype data are used to develop a predictive model that links variation in genetic markers to variation in phenotypes of interest. Training populations are usually derived from existing progeny trials or breeding populations, where previously selected 'elite' parents have been crossed, and their progenies have been extensively tested in field trials. Breeding populations in forest trees usually contain 1000–2000 individuals and have effective population sizes (N_e) in the range of 30–100. The larger the training population, while keeping N_e in the appropriate range, the more precisely marker effects are estimated, ultimately resulting in a more accurate predictive model [88].

GS is then employed by genotyping a large number of individuals that form 'selection candidates', usually consisting of full- or half-sib families derived from individuals that are part of the training population. The genotype information is used with the prediction model to estimate genomics-based genotypic values (GEGVs) for all selection candidate individuals. Top-ranking individuals from the selection candidates, based on the GEGVs, are then selected and used to establish the next generation of the breeding program. To further enhance testing, top-ranked selection candidates can also be clonally propagated and tested in clonal trials, where elite clones are eventually selected for operational plantation, especially for tree species that rely on clonal deployment, such as *Eucalyptus* and *Populus*. A random subset of the selection candidates can also be planted in an experimental design trial and phenotyped at the target age to update the GS model and ensure that the accuracy of GS predictions remains high over successive generations.

Large training population sizes have thus far characterized studies of GS in forest trees, and large numbers of genetic markers have also typically been used, especially when compared to GS studies in crops. Recent studies in forest trees include a diverse array of species, including eucalypts [89, 90], spruces [91, 92], pines [93], and *Populus* [40, 94, 95]. These studies have shown that the ability to predict complex traits in forest trees is high and suitable for implementing GS as a breeding tool to increase the efficiency of tree breeding programs. Furthermore, in species that utilize clonal deployment, such as eucalypts and poplars, GS can reduce or eliminate the progeny trials and the time and costs of clonal testing trials by minimizing the number of selected genotypes propagated as clones.

GS's utility depends on fundamental population and quantitative genetics aspects as well as more practical and logistical aspects relating to resource allocation and cost-benefit considerations. The accuracy of a genomic prediction model is the most important aspect of the success of GS, and many factors immediately impact GS model accuracy. The first important aspect is the effective size of the training population (N_e), which directly influences the extent of linkage disequilibrium (LD), which, in turn, dictates the necessary genetic marker density needed for successful model building. The following essential aspect is phenotyping accuracy in the training population, which sets an upper limit to how much variation can be explained by marker effects. Similarly, the heritability and genetic architecture of the traits of interest are essential, and traits with high heritabilities generally have higher prediction accuracies. Finally, the statistical methods used for model building and deployment are also relevant, and this is currently an area of active research. Simulation studies have partially assessed all of these factors and have provided some general guidelines for implementing GS in forest trees [86, 89].

The extent of LD is perhaps the most important aspect influencing the accuracy of GS, and LD, in turn, depends on the N_e of the training population. The extent of LD directly affects the marker density needed for successful implementation of GS, and marker density has been shown to tightly scale with N_e of the training population, where larger populations need more markers. The level of LD between markers and the unknown quantitative trait loci (QTLs) controlling traits of interest can be increased by reducing N_e. Previous studies have shown that to maintain reasonable levels of LD while maintaining sufficient genetic diversity to sustain long-term genetic gains in breeding, N_e in the range of 40–100 is typically recommended. These N_es typically correspond to census sizes of 100–200 related individuals; for example, most advanced GS populations in *Eucalyptus* have N_es around 30–60 [86, 96]. For such small values of N_e, the accuracy of GS can be maintained with marker densities averaging 2–3 markers per centiMorgan (cM). For a genome of 1500–2000 cM, typical for many species of, e.g., *Eucalyptus* and *Populus*, ~5000 SNPs would be sufficient for most practical applications of GS. However, as N_e increases, the density of markers also must increase to maintain GS accuracy, and up to 20 markers/cM can be needed to maintain accuracy as N_e approaches and exceeds 100–200 individuals [86], requiring genotyping methods that routinely and reliably can genotype 20,000–50,000 informative markers depending on the size of the target genome. Another critical aspect of marker density worth emphasizing is that higher marker densities will facilitate the preservation of rare alleles and thereby contribute to long-term gains from selection in the breeding population [97].

The design of the training population depends on the actual breeding strategy that is adopted. Training populations are established using trees from existing progeny trials based on crosses among elite parents. The relatedness between the training population and selection of candidates is another factor that has great importance for the success of GS. Increasing the genetic relationships between the training population and selection candidates also leads to higher prediction accuracies in a manner similar to the effects of reducing N_e [98]. Similarly, the genetic architecture underlying trait variation has significant consequences for the accuracy of GS. A small number of loci controlling a large proportion of the phenotypic variation allows for more variation to be captured compared to more complex genetic architectures involving more loci, each with a smaller effect size [99]. GS accuracy drops with the increasing number of QTLs contributing to a trait, and this effect is more pronounced when marker density is low or in populations with large N_e [86]. Traits heritability only has minor impacts on GS accuracy as long as the training population size is large enough to estimate marker effects adequately.

Based on the issues outlined in the preceding paragraphs, the prospects for implementing GS in *Populus* are good. There are abundant genomic resources, including large SNP data sets, available in many *Populus* species. What perhaps is lacking are cheap and reliable methods for large-scale genotyping of SNP data sets of the size that are useful for establishing GS (5–50k SNPs). Although SNP chips have been developed for some species of *Populus* [80], no commercial alternatives are available akin to the EuCHIP60K that has been developed in *Eucalyptus* and can be used across several species in the genus [100]. However, several genotyping methods suitable for targeted genotyping have recently started to become commercially available at costs that make them feasible for commercial GS programs. In addition, many species of *Populus* have extensive clone collections available, and many species also have comprehensive progeny testing programs ongoing, meaning that ample material is available for use in breeding programs and in establishing training populations.

9. Conclusions

Extensive collaboration between forestry sector and academia in the Nordic-Baltic region over the past decades has resulted in readiness for successful hybrid poplar breeding programs in the region. The existing collections of *Populus* clones with large genetic variations in adaptive phenology originating from different geographical regions and breeding programs in the northern hemisphere have paved the ground for the establishment of training populations with effective population sizes for successful genomic selection. Deployment of genomic selection models in poplar breeding reduces the duration of breeding cycles and leads to availability of complementary woody raw material for the forestry sector in the Nordic-Baltic region. Poplars are well-suited for genomic selection due to the abundance of genomic resources available, including large panels of genetic markers. Also, target phenology traits, such as growth cessation, have generally high heritabilities suggesting that selection will be efficient. Hybrid poplars as early successional trees grown in short rotations of 10–20 years provide a significant complementary wood source for Nordic-Baltic forest industries that today rely mostly on secondary successional tree species grown in rotations of 50–100 years.

Acknowledgements

The authors will greatly acknowledge our partners from the forestry sector who have been driven in innovations in the procurement of woody raw material in the Nordic-Baltic region: Anders Ekstrand from Södra Skog AB, Lars-Georg Hedlund from Södra Latvia SIA and Mindaugas Šilininkas from Euromediena UAB. This chapter has been written within three following projects during the last decade: (1) an EU project within the EUREKA program Eurostars E! 8443—SnowTiger during 2014–2016, (2) grant number 942-2016-20001 from Swedish Research Council FORMAS "Climate-adapted poplar through more efficient breeding and better tools for matching genotype and site—developing the poplar bio-economy market in Sweden and the Baltic Region" during 2016–2021. Finally, (3) this chapter was finalized within the "NutriBiomass4LIFE" project (LIFE17/ENV/LV000310) co-financed by the Swedish Energy Agency (P-45082-1).

Author details

Anneli Adler[1*], Almir Karacic[1], Rami-Petteri Apuli[2], Ann-Christin Rönnberg Wästljung[2], Magnus Hertzberg[3], Martin Weih[1] and Pär K. Ingvarsson[2]

1 Department of Crop Production Ecology, Swedish University of Agricultural Sciences, Uppsala, Sweden

2 Department of Plant Biology, Swedish University of Agricultural Sciences, Uppsala, Sweden

3 SweTree Technologies AB, Umeå, Sweden

*Address all correspondence to: anneli.adler@slu.se

References

[1] Jonsson TH, Oskarsson U. Shoot growth strategy of 29 black cottonwood (*Populus trichocarpa*) clones. Icelandic Agricultural Sciences. 2007;**20**:25-36

[2] Sverrisson H. A Review of the Icelandic Poplar Breeding Program. Iceland: Iceland Forest Service; 2017

[3] Ilstedt B. Genetics and performance of Belgian poplar clones tested in Sweden. Forest Genetics. 1996;**3**(4):183-195

[4] Ilstedt B. Anpassning av Populus trichocarpa, jättepoppel, till svenskt klimat. In: Proceedings from a Poplar Seminar at the Department of Short Rotation Forestry. Uppsala: Swedish University of Agricultural Sciences; 2005

[5] Rohde A et al. Bud set in poplar—Genetic dissection of a complex trait in natural and hybrid populations. New Phytologist. 2011;**189**(1):106-121

[6] Ronnberg-Wastljung AC et al. Phenotypic plasticity in Populus trichocarpa clones across environments in the Nordic-Baltic region. Scandinavian Journal of Forest Research. 2022;**37**(1):1-5

[7] Adler A et al. Variation of growth and phenology traits in poplars planted in clonal trials in Northern Europe-implications for breeding. Bioenergy Research. 2021;**14**(2):426-444

[8] Karacic A et al. An analysis of poplar growth and quality traits to facilitate identification of climate-adapted plant material for Sweden. Bioenergy Research. 2021;**14**(2):409-425

[9] Farmer REJ. The genecology of Populus. In: Stettler R, Bradshaw HD Jr, Heilman PE, Hinckley TM, editors. Biology of *Populus* and Its Implications for Management and Conservation. NRC Research Press, National Research Council of Canada: Ottawa, ON, Canada; 1996. pp. 33-55

[10] Dickmann DI, Kuzovkina J. Poplars and Willows of the World, with Emphasis on Silviculturally Important Species. Rome, Italy: FAO/IPC; 2008. p. 134. Working Paper IPC/9-2, in FAO/IPC Poplars and Willows in the World

[11] Soolanayakanahally RY et al. Timing of photoperiodic competency causes phenological mismatch in balsam poplar (*Populus balsamifera* L.). Plant Cell and Environment. 2013;**36**(1):116-127

[12] Hetemäki L. The outlook for Nordic-Baltic forest bioeconomy to 2030. In: Liuhto K, editor. The Forest Industry around the Baltic Sea Region:Future Challenges and Opportunities. Helsinki: Centrum Balticum Foundation; 2020. pp. 14-24

[13] Lundmark T. Skogen räcker inte till—hur ska vi prioritera? In Swedish. In: Future Forests Rapportserie 2020. Umeå: U.o.A. Sciences; 2020. p. 24

[14] Schlyter P et al. Assessment of the impacts of climate change and weather extremes on boreal forests in northern Europe, focusing on Norway spruce. Climate Research. 2006;**31**(1):75-84

[15] IPC. Synthesis of country progress reports. The International Commission on poplars and other fast-growing trees sustaining people and the environment (IPC). Twent-sixth session. FO:IPC/2021/Inf2. 2021: FAO, Rome. 2021

[16] Estonian Cell. 2023. About us. Available from: https://www.estoniancell.ee/en/about-us/facts-figures/ [Accessed: September 30, 2023]

[17] Jensen V. Snabbväxande lövträd kan öka avkastning och sprida dina risker. In Swedish. SkogsAktuellt, Independent magazine on forestry in Sweden. 2023;3

[18] FAOSTAT. 2023. Available from: https://www.fao.org/faostat/en/#data/FO [Accessed: August 1, 2023]

[19] Hedlund L.-G. CEO of Södra Latvia SIA, personal communication. 2023

[20] Adler A et al. Country Report 2016-2019 from National Commission of Fast-Growing Trees in Sweden. Uppsala: National Commission of Fast-Growing Trees in Sweden; 2020. p. 18

[21] Xu X, Mola-Yudego B. Where and when are plantations established? Land-use replacement patterns of fast-growing plantations on agricultural land. Biomass & Bioenergy. 2021;**144**:105921. pp. 1-10

[22] Pra A, Pettenella D. Investment returns from hybrid poplar plantations in northern Italy between 2001 and 2016: Are we losing a bio-based segment of the primary economy? Rivista di Economia Agraria. 2019;**74**(1):49-71

[23] Hungate BA et al. The economic value of grassland species for carbon storage. Science Advances. 2017;**3**(4):e1601880. p. 8

[24] Stener L-G, Rytter L, Beuker E, Tullus H, Lutter R. Hybrid Aspen and Poplars in the Baltic Sea Region and Iceland—Results from a Questionnaire and a Literature Review. Uppsala, Sweden: Skogforsk; 2019

[25] Adler A et al. Lignin-first biorefining of Nordic poplar to produce cellulose fibers could displace cotton production on agricultural lands. Joule. 2022;**6**(8):1845-1858

[26] Eckenwalder J. Systematics an evolution of Populus. In: Stettler R, Bradshaw HD Jr, Heilman PE, Hinckley TM, editors. Biology of Populus and its Implications for Management and Conservation. Ottawa, ON, Canada: NRC Research Press, National Research Council of Canada; 1996. pp. 7-32

[27] Little EL Jr. Atlas of United States Trees, Vol. 1, Conifers and Important Hardwoods: U.S. Department of Agriculture Miscellaneous Publication No. 1146, 200 maps. Public domain to the map via Wikimedia Commons. 1971. 9 p. Available from: https://upload.wikimedia.org/wikipedia/commons/d/da/Populus_trichocarpa_range_map_1.png

[28] Dickmann DI. An overview of genus Populus. In: Dickmann DI, Isebrands JG, Eckenwalder JE, Richardson J, editors. Poplar Culture in North America. Ottawa, ON K1A 0R6, Canada: NCR Press, National Research Council of Canada; 2001. pp. 1-42

[29] Silininkas M. Personal communication with Mindaugas Silininkas, CEO Euromediena UAB. 2023

[30] Stener LG. Tillväxt, vitalitet och densitet för kloner av hybridasp och poppel i sydsvenska fältförsök, in Arbetsrapport från SkogForsk nr 717. Uppsala. 2010; 50 pp

[31] Elferjani R, DesRochers A, Tremblay F. Effects of mixing clones on hybrid poplar productivity, photosynthesis and root development in northeastern Canadian plantations. Forest Ecology and Management. 2014;**327**:157-166

[32] Karacic A, Verwijst T, Weih M. Above-ground woody biomass production of short-rotation populus plantations on agricultural land in

Sweden. Scandinavian Journal of Forest Research. 2003;**18**(5):427-437

[33] Jónsson B. Coppice silviculture of black cottonwood for production of wood chips—Effects of harvesting date on regeneration and yield. In: The WoodBio Conference. Reykjavik, Iceland; 2017

[34] Johnsson A. Föreningen för växtförädling av skogsträd—1936-1958. In Swedish, F.f.v.a. skogsträd, Editor. Uppsala, Sweden. 1959

[35] Eriksson H. Yield of aspen and poplars in Sweden. In: Perttu K, editor. Ecology and Management of Forest Biomass Production Systems. Swedish: Department of Ecology and Environmental Research, Swedish University of Agricultural Sciences; 1984. pp. 393-419

[36] Karlsson B, Werner M, Stener LG. Resultat från två klonförsök med poppel, in Arbetsrapport från SkogForsk nr 319. Uppsala. 1996. 16 p

[37] Stener LG. Resultat från sydsvenska klontester med poppel. Skogforsk. 2004

[38] Stener LG, Westin J. Early growth and phenology of hybrid aspen and poplar in clonal field tests in Scandinavia. Silva Fennica. 2017;**51**(3):5656. p. 22

[39] Christersson L. Biomass production of intensively grown poplars in the southernmost part of Sweden: Observations of characters, traits and growth potential. Biomass & Bioenergy. 2006;**30**(6):497-508

[40] Stanton BJ, Neale DB, Li S. Populus breeding: From the classical to the genomic approach. In: Jansson S, Bhalerao RP, Groover AT, editors. Genetics and Genomics of Populus. Vol. 8. 2010. pp. 309-348

[41] Steenackers MS, de Uyper C, Michiels B. Breeding and selection of poplars for durable resistance to Melampsora larici-Populina. In: First IUFRO Rusts of Forest Trees Working Party Conference, August 2-7 1998. Saariselkä, Finland: Finnish Forest Research Institute, Rovaniemi Research Station; 1998

[42] Ilstedt B. Breeding strategy for poplar Populus trichocarpa, jättepoppel, till svenskt klimat. Norwegian Journal of Agricultural Sciences. 1994:39-45

[43] Richards TJ et al. Quantitative genetic architecture of adaptive phenology traits in the deciduous tree, *Populus trichocarpa* (Torr. and Gray). Heredity. 2020;**125**(6):449-458

[44] Vico G et al. Consistent poplar clone ranking based on leaf phenology and temperature along a latitudinal and climatic gradient in Northern Europe. Bioenergy Research. 2021;**14**(2):445-459

[45] Olson MS et al. The adaptive potential of *Populus balsamifera* L. to phenology requirements in a warmer global climate. Molecular Ecology. 2013;**22**(5):1214-1230

[46] Fabbrini F et al. Phenotypic plasticity, QTL mapping and genomic characterization of bud set in black poplar. BMC Plant Biology. 2012;**12**:47. p. 16

[47] Ducros ET. Breeding strategies with poplars in EUROPE. Forest Ecology and Management. 1984;**8**(1):23-39

[48] Steenackers J et al. Poplar diseases, consequences on growth and wood quality. Biomass & Bioenergy. 1996;**10**(5-6):267-274

[49] Swedish Meteorological and Hydrological Institute (SMHI). 2017.

Available from: https://www.smhi.se/data/meteorologi/temperatur/genomsnittliga-datum-for-den-forsta-hostfrosten-1.4074 [Accessed: August 5, 2023]

[50] Frewen BE et al. Quantitative trait loci and candidate gene mapping of bud set and bud flush in *Populus*. Genetics. 2000;**154**(2):837-845

[51] Rohde A et al. PtABI3 impinges on the growth and differentiation of embryonic leaves during bud set in poplar. Plant Cell. 2002;**14**(8):1885-1901

[52] Bohlenius H et al. CO/FT regulatory module controls timing of flowering and seasonal growth cessation in trees. Science. 2006;**312**(5776):1040-1043

[53] Ingvarsson PK et al. Nucleotide polymoirphism and phenotypic associations within and around the phytochrome B2 locus in European aspen (*Populus tremula*, Salicaceae). Genetics. 2008;**178**(4):2217-2226

[54] Ruonala R et al. CENL1 expression in the rib meristem affects stem elongation and the transition to dormancy in *Populus*. Plant Cell. 2008;**20**(1):59-74

[55] Jimenez S et al. Phylogenetic analysis and molecular evolution of the dormancy associated MADS-box genes from peach. BMC Plant Biology. 2009;**9**:81. p. 12

[56] Li Z et al. Dormancy-associated MADS genes from the EVG locus of peach *Prunus persica* (L.) Batsch have distinct seasonal and photoperiodic expression patterns. Journal of Experimental Botany. 2009;**60**(12):3521-3530

[57] Ibanez C et al. Circadian clock components regulate entry and affect exit of seasonal dormancy as well as winter hardiness in *Populus* trees. Plant Physiology. 2010;**153**(4):1823-1833

[58] Rohde A, Bastien C, Boerjan W. Temperature signals contribute to the timing of photoperiodic growth cessation and bud set in poplar. Tree Physiology. 2011;**31**(5):472-482

[59] Ruttink T et al. A molecular timetable for apical bud formation and dormancy induction in poplar. Plant Cell. 2007;**19**(8):2370-2390

[60] Cooke JEK, Eriksson ME, Junttila O. The dynamic nature of bud dormancy in trees: Environmental control and molecular mechanisms. Plant Cell and Environment. 2012;**35**(10):1707-1728

[61] Savolainen O, Lascoux M, Merila J. Ecological genomics of local adaptation. Nature Reviews Genetics. 2013;**14**(11):807-820

[62] Olsen JE. Light and temperature sensing and signaling in induction of bud dormancy in woody plants. Plant Molecular Biology. 2010;**73**(1-2):37-47

[63] Evans LM et al. Population genomics of *Populus trichocarpa* identifies signatures of selection and adaptive trait associations. Nature Genetics. 2014;**46**(10):1089-1096

[64] McKown AD et al. Genome-wide association implicates numerous genes underlying ecological trait variation in natural populations of *Populus trichocarpa*. New Phytologist. 2014;**203**(2):535-553

[65] Apuli R-P et al. The genetic basis of adaptation in phenology in an introduced population of Black Cottonwood (*Populus trichocarpa*, Torr. & Gray). BMC Plant Biology. 2021;**21**(1):317

[66] McWatters HG, Devlin PF. Timing in plants—A rhythmic arrangement. FEBS Letters. 2011;**585**(10):1474-1484

[67] Howe GT et al. Evidence that the phytochrome gene family in black cottonwood has one PHYA locus and two PHYB loci but lacks members of the PHYC/F and PHYE subfamilies. Molecular Biology and Evolution. 1998;**15**(2):160-175

[68] Olsen JE et al. Ectopic expression of oat phytochrome A in hybrid aspen changes critical daylength for growth and prevents cold acclimatization. Plant Journal. 1997;**12**(6):1339-1350

[69] Hall D et al. Adaptive population differentiation in phenology across a latitudinal gradient in European Aspen (*Populus tremula*, L.): A comparison of neutral markers, candidate genes and phenotypic traits. Evolution. 2007;**61**(12):2849-2860

[70] Wang J et al. A major locus controls local adaptation and adaptive life history variation in a perennial plant. Genome Biology. 2018;**19**:72

[71] Zhang M, Suren H, Holliday JA. Phenotypic and genomic local adaptation across latitude and altitude in *Populus trichocarpa*. Genome Biology and Evolution. 2019;**11**(8):2256-2272

[72] Jansson S, Douglas CJ. *Populus*: A model system for plant biology. Annual Review of Plant Biology. 2007;**58**: 435-458

[73] Bradshaw HD et al. Emerging model systems in plant biology: Poplar (*Populus*) as a model forest tree. Journal of Plant Growth Regulation. 2000;**19**(3):306-313

[74] Tuskan GA et al. The Genome of Black Cottonwood, *Populus trichocarpa* (Torr. & Gray). Science. 2006;**313**(5793): 1596-1604

[75] Ingvarsson PK. Nucleotide polymorphism and linkage disequilbrium within and among natural populations of European Aspen (*Populus tremula* L., Salicaceae). Genetics. 2005;**169**(2):945-953

[76] Kelleher CT et al. A physical map of the highly heterozygous *Populus* genome: Integration with the genome sequence and genetic map and analysis of haplotype variation. Plant Journal. 2007;**50**(6):1063-1078

[77] Olson MS et al. Nucleotide diversity and linkage disequilibrium in balsam poplar (*Populus balsamifera*). New Phytologist. 2010;**186**(2):526-536

[78] Slavov GT et al. Genome resequencing reveals multiscale geographic structure and extensive linkage disequilibrium in the forest tree *Populus trichocarpa*. New Phytologist. 2012;**196**(3):713-725

[79] Wang J et al. Natural selection and recombination rate variation shape nucleotide polymorphism across the genomes of three related *Populus* species. Genetics. 2016;**202**(3):1185

[80] Geraldes A et al. A 34K SNP genotyping array for *Populus trichocarpa*: Design, application to the study of natural populations and transferability to other *Populus* species. Molecular Ecology Resources. 2013;**13**(2):306-323

[81] Schilling MP et al. Genotyping-by-sequencing for *Populus* population genomics: An assessment of genome sampling patterns and filtering approaches. PLoS One. 2014;**9**(4):e95292

[82] Faivre-Rampant P et al. New resources for genetic studies in *Populus nigra*: Genome-wide SNP discovery

and development of a 12k Infinium array. Molecular Ecology Resources. 2016;**16**(4):1023-1036

[83] Isabel N, Lamothe M, Thompson SL. A second-generation diagnostic single nucleotide polymorphism (SNP)-based assay, optimized to distinguish among eight poplar (*Populus L.*) species and their early hybrids. Tree Genetics & Genomes. 2013;**9**(2):621-626

[84] Chhetri HB et al. Multitrait genome-wide association analysis of *Populus trichocarpa* identifies key polymorphisms controlling morphological and physiological traits. New Phytologist. 2019;**223**(1):293-309

[85] Meuwissen THE, Luan T, Woolliams JA. The unified approach to the use of genomic and pedigree information in genomic evaluations revisited. Journal of Animal Breeding and Genetics. 2011;**128**(6):429-439

[86] Grattapaglia D, Resende MDV. Genomic selection in forest tree breeding. Tree Genetics & Genomes. 2011;7(2):241-255

[87] Nakaya A, Isobe SN. Will genomic selection be a practical method for plant breeding? Annals of Botany. 2012;**110**(6):1303-1316

[88] Hayes BJ et al. Accuracy of genomic selection: Comparing theory and results. Matching genetics and environment: a new look at an old topic. In: Proceedings of the 18th Conference of the Association for the Advancement of Animal Breeding and Genetics, September 28–October 1, 2009. Barossa Valley, South Australia, Australia; 2009. pp. 34-37

[89] Resende MDV et al. Genomic selection for growth and wood quality in Eucalyptus: Capturing the missing heritability and accelerating breeding for complex traits in forest trees. New Phytologist. 2012;**194**(1):116-128

[90] Tan B et al. Evaluating the accuracy of genomic prediction of growth and wood traits in two Eucalyptus species and their F-1 hybrids. BMC Plant Biology. 2017;**17**:110

[91] El-Dien OG et al. Prediction accuracies for growth and wood attributes of interior spruce in space using genotyping-by-sequencing. BMC Genomics. 2015;**16**:370

[92] Ratcliffe B et al. A comparison of genomic selection models across time in interior spruce (Picea engelmannii x glauca) using unordered SNP imputation methods. Heredity. 2015;**115**(6):547-555

[93] de Almeida Filho JE et al. The contribution of dominance to phenotype prediction in a pine breeding and simulated population. Heredity. 2016;**117**(1):33-41

[94] Kirst M, Resende M, Munoz P, Neves L. Capturing and genotyping the genome-wide genetic diversity of trees for association mapping. In: BMC Proceedings, IUFRO Tree Biotechnology Conference 2011: From Genomes to Integration and Delivery, Arraial d'Ajuda, Bahia, Brazil. 2011. pp. 1-2

[95] Alves FC et al. Accelerating forest tree breeding by integrating genomic selection and greenhouse phenotyping. Plant Genome. 2020;**13**(3):e20048

[96] Kerr RJ et al. Simulation of hybrid forest tree breeding strategies. Canadian Journal of Forest Research. 2004;**34**(1):195-208

[97] Perez-Enciso M, Rincon JC, Legarra A. Sequence- vs. chip-assisted genomic selection: Accurate biological

information is advised. Genetics Selection Evolution. 2015;**47**:43

[98] Beaulieu J et al. Genomic selection accuracies within and between environments and small breeding groups in white spruce. BMC Genomics. 2014;**15**:1048

[99] Grattapaglia D, do Amaral Diener PS, dos Santos GA. Performance of microsatellites for parentage assignment following mass controlled pollination in a clonal seed orchard of loblolly pine (*Pinus taeda* L.). Tree Genetics & Genomes. 2014;**10**(6):1631-1643

[100] Silva-Junior OB, Grattapaglia D. Genome-wide patterns of recombination, linkage disequilibrium and nucleotide diversity from pooled resequencing and single nucleotide polymorphism genotyping unlock the evolutionary history of Eucalyptus grandis. New Phytologist. 2015;**208**(3):830-845

Chapter 4

Stress Memory and Priming Applications in Plants: Potential for Breeders

Tamer Gümüş, Tuğçe Aydın, Burcu Gündüz, Sinan Meriç, Alp Ayan and Çimen Atak

Abstract

Humankind interfered in the natural selection of plants in favor of traits such as yield, grain quality, productivity, and flavor principally at the expense of several biotic and abiotic stress tolerance capacities. Plants are subjected to the detrimental effects of the combination of these factors due to their stationary nature. Today, there are various breeding approaches from classical to transgenesis and even genome editing to tame plant genome for our purposes. Additionally, the significance of epigenetic regulation in response to biotic and abiotic stresses has been recognized in the last decade. Acquisition and preservation of stress memory for the progeny to allow them to adapt to similar conditions through methylation, histone modification, and chromatin structure alterations are the focus of attention. Enlightening the cross talk between these components of acquired transgenerational memory may aid to breed more efficient and environmentally friendly crops in current agricultural systems. Priming applications have been extensively studied to induce stress memory of the plant by external stimulus as a warning signal, which may ignite minor activations of stress-responsive gene expression and eventually turn into strong resistance. The present chapter will discuss the basis and the recent advances in plant epigenetic regulation with emphasis on chemical, biotic, and abiotic priming agents.

Keywords: epigenetics, epigenomics, histone modifications, methylation, plant breeding

1. Introduction

Crop production and global food security are endangered due to the rapid increase in the world population and the continuous emergence of drought, flood, heat wave, and frost events along with other global climate change-related issues. Therefore, new plant breeding strategies are required for achieving stress tolerance, increasing yield, improving crop quality, and creating more adaptive and sustainable germplasms for future climate challenges. In recent years, transgenic technology and genome editing techniques, which include the transformation and genetic modification of single or multiple genes, as well as classical or mutation breeding, have been used efficiently for stress-tolerant or -resistant plant breeding. Clearly, all breeding methods have advantages and disadvantages for the improvement of agricultural traits. Classical breeding and hybridization are the most common and traditional methods for environmental stress-tolerant varieties. However, the breeding process depends on the availability of the germplasm that has the desired traits. The breeding process takes a very long time throughout generations. Traits are only inherited between closely related species. Selection of the desired traits in candidate lines requires intensive labor. All these limitations led plant breeders to search for new strategies as alternatives to classical breeding. In mutation breeding, various physical, chemical, or biological mutagens are used to increase genetic variability. High mutation frequency is obtained compared to naturally occurring spontaneous mutations [1–4]. The selection process requirement in each generation is a major disadvantage of mutation breeding. On the other hand, transgenic technology has revolutionized as the fastest adopted crop technology in the history of modern agriculture, which enables the improvement of plants with the predictable changes for target trait in a relatively short time [5–8]. However, the restrictive effects of legal regulations regarding the cultivation of genetically modified crops and their transport into other countries are some of the obstacles in front of this strategy as well as metabolic imbalance or off-target effects caused by genetic modification [9, 10]. All these difficulties encountered in various plant breeding strategies and the radical expansion of the genetic and epigenetic information pool have paved the way for the testing of novel strategies.

Since plants have a stationary lifestyle and lack adaptive immune system as well as specific immune response cells, they require highly sophisticated regulatory mechanisms for defense. Memorizing a past stressor and the proper response is extremely cost-effective to plants compared to constitutive preparedness. Since plants lack the nervous system and a central brain, retrieving the previous response to a specific stressor after a period of unstressed condition may delay effective counteract. In this regard, various priming applications are getting tested today to put plants into a physiological state that allows them to respond more rapidly and/or more robustly after exposure to a particular biotic or abiotic stress. In this state, plants have enhanced stress perception due to the amplified stress signaling that leads to more efficient activation of the defense response and enhanced resistance before encountering the stressor, instead of stress response-related gene induction [11]. Stress-induced memory is mediated by metabolomic (amino acids, sugars, etc.), proteomic (antioxidant and photosynthesis enzymes etc.), transcriptomic (WRKY, AREB, etc.), epigenetic (DNA methylation, chromatin remodeling, etc.) mechanisms. These mechanisms can be transferred to future generations [12]. Among these mechanisms, epigenetics particularly play an important role in creating, preserving, and transmitting stress memory to the next generations. Some stress-induced epigenetic changes can act as a transmittable memory for the progeny. For these reasons, it is important to comprehend three main mechanisms of epigenetic regulation to understand stress memory and the basis of priming applications: DNA methylation, histone modifications, and RNA-mediated gene silencing.

In the last decade, epigenetic regulatory mechanism of gene function without alteration of the DNA sequence was an intense research topic. In 1994, Wassenegger and colleagues discovered induction of *de novo* 5mC methylation of genes under viroid infection of transgenic tobacco plants for the first time. They assumed that specific methylation serves to protect against overexpression of primary viroid cDNA [13]. Today, the role of DNA methylation is not only verified but also an intensively studied topic with much yet to be discovered as suggested. DNA methylation has relatively stable marks on plant genome under certain conditions, which are also heritable and transgenerational. These marks are related to both environmental stress responses and regular physiological processes as temporal and spatial gene expression, as well as transposon mobility, and genomic imprinting [14]. DNA methylation mostly involves the addition of methyl group onto the C5 position of the cytosine to form 5-methyl-cytosine (5mC). It can occur in promoter regions as well as protein-coding regions of genes. In plants, it can also occur in all cytosine sequence contexts as CG, CHG, and HH (where H represents A, T, or C), while in mammals, it is mainly CG methylation. Other ubiquitously detected plant methylation types are tissue and development stage specific in mammal cells. After the addition of methyl group, methyltransferases are primarily responsible for the maintenance of methylome during replication through cell division. Besides the maintenance, *de novo* DNA methylation, which is methylation at a previously unmethylated C and DNA demethylation processes, may occur. DNA demethylation can occur passively by the reduction or inactivation of key enzymes in DNA methylation process or actively in much complicated mechanism through the involvement of base excision repair (BER) pathway [15].

Histones are the central components of chromatin organization. There is a group of histones that are predominantly dependent on DNA replication known as canonical histones and also a group called replacement histones that are expressed all along the cell cycle and independent of DNA replication. They are localized in specific regions in the genome and have their unique sequences among themselves. Replacement histones, also known as histone variants, are found in all studied model organisms. Some of them can be lineage-, tissue-, or domain-specific. For instance, H2A.X and H2A.Z are evolutionarily conserved from yeast to mammals, whereas macroH2A and H2A.Bbd are found only in mammals and H2A.W is specific to plants [16]. Eukaryotic genomes consist of repeating nucleosomes that are octamer histones wrapped around 147 base pairs of DNA. Nucleosomes that were previously believed to provide a universal, nonspecific coating of genomic DNA are well-known to have favored positions throughout the genome [17]. Depending on the nucleosome composition, chromatin either can be more accessible to transcription or has closed conformation. The different states of chromatin are constituted from diverse canonical and replacement histone combinations. H3.3, H2A.Z, and H2A.X variants are abundant in euchromatic regions. H3.3 evolved independently in plants and animals. The evidence of convergent evolution strongly points toward the importance of H3.3 to the function of the eukaryotic genome [18]. H2A.Z which has a single evolutionary origin nearly in branches of Eukarya. It plays pivotal roles in multicellular development. It has been linked to various biological (plant immunity, germline development, and stress response) and cellular (genome stability and DNA repair) processes as well as both transcriptional activation and repression. H2A.X is found in most of the eukaryotes; however, it has evolved multiple times contrary to the H2A.Z. Phosphorylated H2A.X is suggested to be a histone mark for DNA damage repair (DDR) [18]. There are other histone marks in chromatin state typical of active transcription. H3K4me3 is particularly enriched at the transcriptional start site of actively transcribed genes. H3K36me3 differs from that in animals as it marks active genes in plants at 5′ region similar to H3K4me3. It is related to gene body of active genes. H2B can be monoubiquitinated at their C-terminal residues (H2Bub1). This mark is usually related to active genes and associated to genomic regions in both H3K4me3 and H3K36me3 [19]. On the other hand, there are also heterochromatin marks in silent genomic regions allowing chromatin compaction such as H3K9me2, H3K27me3, H3K27me1 [20]. During stress conditions, some histone variants are incorporated in the nucleosomes of stress-responsive genes. They may have various posttranscriptional modifications (PTMs) such as acetylation, methylation, and phosphorylation. The histone marks and their PTMs allow a wide variety of nucleosome combinations. Therefore, they are responsible for storing plant epigenetic memory under different circumstances [21].

Small RNAs (such as micro RNAs and small interfering RNAs) play an important role in regulating gene expression through transcriptional gene silencing (TGS) or posttranscriptional gene silencing (PTGS) under environmental stress. Stress-induced synthesis of small RNAs can reduce the expression of genes that would be affected by stress conditions. Conversely, they can downregulate their own expression to increase the expression of target genes. There have been many studies showing the relationship between stress memory and small RNAs [22, 23]. The chromatin remodeling machinery, which may act alone or in complexes, is controlled by CHR (chromatin remodeling factors) proteins such as histone chaperones, histone-modifying enzymes, and ATP-dependent chromatin remodeling complexes. Chromatin remodeling compresses and unfolds nucleosome DNA. Therefore, DNA becomes accessible for replication, repair, and selective gene expression. In this way, it plays a role as an important mechanism in the biotic and/or abiotic stress response in plants. Numerous evidence presented the relationship between stress memory and chromatin remodeling [24–26].

The existence of epigenetic memory within the generation is called intragenerational adaptation. Most of the epigenetic alterations are reversible after prolonged unstressed conditions; however, subsequent exposures may lead to stress training. It has been well-documented particularly for drought stress memory [27, 28]. It is also known that some of these trained chromatin alterations are heritable by meiosis. They are called transgenerational inheritance of stress memory. They can be detected at least after a non-stressed generation of a parental plant that has experienced the environmental effect. Otherwise, if subsequent progeny does not present the trained responses, it is simply intragenerational adaptive phenotypic change in parents.

2. Stress memory in plants

Stress is a state where environmental conditions change enough to affect the normal growth and development of a plant adversely. Therefore, it leads to a wide range of molecular alterations throughout the organism. These effects can occur through living organisms (biotic) or environmental factors (abiotic). These stresses may occur unexpectedly or in a cyclical manner over days, seasons, or even years. Some stresses rarely occur throughout a plant's life, but abiotic stresses such as temperature, drought, salinity, and cold are often repetitive. Repeated exposure to different types of stress and re-exposure to a previous stress can alter the biochemical, physiological, transcriptomic, proteomic [29], and metabolomic responses to subsequent stress. Unlike animals, plants are immobile organisms that do not have a specialized nervous system to store the "memories" of past stress events. Stress memory is defined as the capacity of plants to store this information in order to respond to recurring stress in the most rapid manner [30].

The term of "stress memory" in plants was first introduced in 90s, when it was observed that some plants developed resistance to subsequent pathogen attacks after being exposed to a previous pathogen attack [31]. Subsequently, advanced studies have shown that plants exhibit stronger responses to recurring stresses and display much higher tolerance capacities compared to those that have not previously encountered the stress [32].

The concept of stress memory has been categorized under three definitions by Lämke and Bäurle [33]. Somatic stress memory is observed throughout the lifetime of the organism exposed to stress and is mitotically inherited. Somatic stress memory is controlled through mechanisms such as DNA methylation, histone tail modifications, RNA polymerase II pausing, and the production of stress-related metabolites. Even after the stress factor is removed, mechanisms at the DNA, RNA, protein, and chromosome levels, such as gene expression, protein accumulation, and chromatin modifications, remain active [34]. Transgenerational memory (TSM) can be observed even after at least two stress-free generations and is meiotically inherited. It is based on epigenetic foundations. TSM is a phenomenon that surpasses both the offspring's genotype and the direct environmental conditions' impact, due to the transfer of information from stressed parents. TSM can involve the transmission of structural variations in the genome [35], the inheritance of chromatin modifications, or the storage of molecules as maternal-mediated mRNA, hormones, proteins, starch, lipids, and other compounds. Intergenerational memory (ISM) is observed in the first generation of offspring where stress was not encountered. ISM is a memory effect that can only be observed in the first generation where stress is absent. This phenomenon is mediated through the conditions of seed growth and signals produced by the parent plants in the seeds or embryos. The stress factor experienced in the previous generation also affects the reproductive cells and, consequently, the resulting seeds. Thanks to this effect, offspring exhibits phenotypes that cannot otherwise be explained, genotypically [36]. By studying the preconditioning of *Arabidopsis* for hyperosmotic stress, it has been demonstrated that subjecting plant generations to minimal stress during consecutive vegetative stages over two generations can establish this memory. However, after just one stress-free generation, this memory was reset, indicating that adaptation to the environment rapidly diminishes in the absence of stress (**Table 1**).

Plant species	Priming	Stress	Stress memory	Determination of stress memory	Ref.
Arabidopsis thaliana	S-methyl	*Pseudomonas sygringae*	Somatic	Histone methylation level measurement by chromatin immunoprecipitation	[37]
Arabidopsis thaliana	Heat 37°C	Heat 44°C	Somatic	Immunoprecipitation and mass spectrometry	[38]
Arabidopsis thaliana	Salinity	Salinity	Intergenerational	DNA methylation level measurement by bisulfite sequencing	[39]
Solanum tuberosum	β-Aminobutyric acid (BABA)	*Phytophthora infestans*	Intergenerational	DNA methylation level measurement by 5mC ELISA	[40]
Arabidopsis thaliana	Heat 30°C Cold 16°C	Heat 30°C Cold 16°C	Transgenerational	The number of flowers per plant	[41]
Hordeum vulgare	Drought	Drought	Transgenerational	Measurements of photosynthesis and leaf conductance	[42]
Brassica napus	Drought	Drought	Transgenerational	Seed oil content, seed protein content, and seed metabolite content	[43]

Table 1.
Stress memory in various plant species.

Molecular mechanisms related to stress memory are categorized into two main types as cis and trans. The term "cis" refers to the effects of a molecule or genetic material on itself, meaning that cis memory describes changes that occur within the same DNA molecule or genetic region. Cis memory mechanisms are directly influenced by physical markers such as DNA methylation or modifications to histone tails. On the other hand, "trans" memory refers to the effects of a molecule or genetic material on another gene region or molecule. Trans memory involves interactions that are controlled by feedback loops and transmitted to the offspring through cytoplasmic division. It is often used in prokaryotes and single-celled eukaryotes. By another definition, cis memory describes genetic changes that occur within the same gene or DNA region, while trans memory pertains to interactions between different regions of genetic material or molecules. These terms are used to explain processes of genetic-level information storage, transmission, and regulation [34].

Epigenetic mechanisms, including DNA methylation, modification of histone tails, noncoding RNAs, and suppression of 5-serine phosphorylation of RNA polymerase II, play a role in the formation of stress memory. However, besides the epigenetic mechanisms, other mechanisms based on the accumulation of dephosphorylated stress proteins activated by protein kinases, the production of signaling molecules involved in stress perception and tolerance, and changing the levels of transcription factors are also effective. The metabolite accumulation mechanism, which is more temporary, causes short-term stress effects in plants, unlike the long-term stress effects caused by epigenetic mechanisms [44].

Ali et al. reported that the effects of herbivore-induced plant volatiles (HIPV) were memorized and stored by plants. The plant recalls this memory later under secondary herbivore attacks. This epigenetically regulated memory is preserved for at least 5 days after exposure to HIPVs. Moreover, it is reported that serious methylation losses occurred in the promoter region of the Bowman-Birk-type trypsin inhibitor (TI) gene in *Zea mays* with recurrent *Mythimna separata* pest stress [45]. It has been determined that the Flowering Locus C (FLC) gene in *Arabidopsis thalina* changes from a euchromatin structure to a heterochromatin structure under the influence of cold. Chromatin reorganization is regulated from H3K36me3-rich chromatin to H3K27me3-rich chromatin. It has been shown that this is maintained even during periods of rising temperatures [46]. In *Arabidopsis* plants exposed to repeated heat stress, the FORGETTER1 (FGT1) gene, which belongs to a specific heat-inducible gene class, has been reported to participate in stress memory. FGT1 protein binds to the promoter regions of genes involved in heat tolerance, helping these genes to remain accessible to the polymerase even after the removal of the stress [38].

Priming contributes to stronger reactions and long-term survival against stress. This is very important to protect vital activities of the plants. Nevertheless, memory can also prevent the plant's optimal development process. It requires intense energy to keep short-term (physiological, metabolic changes) and long-term (epigenetic) mechanisms ready for stress. Remaining in alerted state even under unstressed conditions may cause disruption of processes such as plant development and reproduction or increase sensitivity to harmful effects. Therefore, removing changes related to stress memory, especially in variable and unpredictable conditions, provides an advantage to the plant. The balance between this state and memory is called "memory resetting". Resetting is established by RNA metabolism, posttranscriptional gene silencing, or RNA-directed DNA methylation [47, 48]. DNA methylation in eukaryotic genomes occurs by the addition of a methyl group to the 5th position of the pyrimidine ring of the cytosine base. While cytosine methylation can occur in the

CG regions in other eukaryotes, cytosine methylation can occur in the CG, CHG, and CHH regions in plant genomes. This difference provides greater epigenetic diversity to plants. Another necessity of memory resetting is due to the density of these methylation regions. As a result of exposure to different stresses, cytosine methylation accumulates at a high rate because of memory formation. The accumulation of methylated cytosine density causes frequent cytosine-thymine exchange that may lead to the enhanced mutation frequency, if not reset [49].

Epigenetic variations in DNA and chromatin are extremely important for the formation of heterochromatin and euchromatin structures. DNA is kept in euchromatin state to respond to different exposed stress factors. This state maintains that the structure highly accessible to physical and chemical mutagens. However, DNA in heterochromatin structure is more protected from mutations. In this regard, euchromatinized DNA regions must be reduced by resetting to protect genome from mutations. Similarly, stress memory stored in the form of proteins must be reset over time. Cytosolic accumulation of proteins affects the functioning of cellular activities. Therefore, protein degradation often occurs through 26S proteasome or autophagy-related ubiquitination [50, 51].

Epigenetic changes that occur randomly or as a result of cellular stress can be inherited across successive generations if not corrected. After stress-induced DNA damage and the random loss of methylation in actively transcribing alleles after DNA damage, the replacement of methylated cytosines with unmethylated cytosines occurs through DNA repair mechanisms. The accumulation of these erroneous epigenetic marks across generations poses a significant problem for the organism. Epigenetic reprogramming mechanisms encompass chromatin remodeling factors and small noncoding RNAs (sncRNAs). They function during gametogenesis and early embryo development to prevent the transmission of acquired chromatin states.

In *Arabidopsis*, stress memory is not transferred equally by female and male germ lines [39]. Most of the stress-related methylation changes detected in leaves were not found in male gametes. Accordingly, stress memory is also transmitted through female germ lines under hyperosmotic conditions. Epigenetic marks caused by environmental factors in male germ lines were efficiently reset by DNA glycosylase (DME), unlike in female gametes. Its reason differs from the meiotic transmission of DNA methylation processes [39]. Female gametes are produced by the mother plant, where they are fertilized and form the seed. Seeds can be distributed to very distant locations. However, they rarely find the opportunity to grow over short distances. Male gametes may have been transported from long distances by wind and animals. Therefore, transmitting transferred stress memory through female germ lines is a more effective way to ensure adaptive responses to stress. The distant male ancestor may have developed memory for completely different stressors under different environmental conditions, but due to their fixed nature, the offspring are likely to be exposed to similar stressful conditions as the female parent [52].

3. Priming applications

The term priming was first introduced in biotic stress studies to ensure immunity against pathogens [53]. Priming involves pre-exposure of plant to an abiotic or biotic stressor to develop resistance. It can be applied at any developmental stage to improve stress tolerance. In addition to natural stressors, it can also be achieved through the application of synthetic stimulants that activate the plant defense systems

[54, 55]. DNA methylation level alterations, histone tail modifications, transcription factors, and inactive mitogen-activated protein kinase accumulation are some important mechanisms of stress memory. These mechanisms are responsible for the re-emergence of stress responses in a much stronger and shorter period in the event of reoccurrence of stress [56].

Priming efficiency in inducing stress memory may vary depending on the preferred priming agent, application time, and plant species (**Table 2**). Therefore, a specific and optimized priming technique is required for each plant species. This is the primary disadvantage of priming, because it requires an extensive number of tests to determine the optimal dose and time for each plant type. There is also a need to develop storage strategies for preservation of primed seeds [64–66]. Optimization involves various parameters such as time required for treatment, priming or coating agent, seed viability, and storage conditions (temperature, humidity, oxygen requirement, etc.), which are standardized for each variety by trial [67]. Priming applications are usually carried out in two different ways for the formation of stress memory. They are seed priming and plant priming. It is an impressive and promising strategy to mitigate the impact of climate change on crops and improve agricultural traits, although still extensive work is required before its application to breeding programs in the field (**Figure 1**) [12, 68].

3.1 Seed priming

Prior to germination, seeds are treated with natural and synthetic components to trigger physiological processes in plants. Nowadays, different priming techniques such as hydropriming, osmopriming, matrix priming, chemical priming, nutripriming, biopriming, and nanoparticle priming or coating seeds with nanoparticles (NPs) are being developed to provide better seed quality, strengthen seeds, accelerate the germination process, and minimize environmental stress [69]. Seed priming plays a role in the formation of stress response through the accumulation of osmolytes (proline, glycine-betaine, and polyamines) by affecting various biochemical processes (cell repair mechanism, antioxidant defense mechanism, etc.) [65].

3.1.1 Hydropriming

Hydropriming is the process of immersing seeds in water to make them stress-tolerant and then drying them to stop germination. This technique is cost-effective compared to other preparation techniques and does not cause toxicity and pollution problems for the seeds and the environment (**Table 3**). The duration of hydropriming is determined by the type of treated seed [66]. It was reported that seedling dry weight and germination percentage increased in hydroprimed basil (*Ocimum basilicum*) plants under salt stress [64]. The most basic approach to ensure hydration is to soak the seed in water and dry it before sowing. Although, unequal hydration and nonuniform germination are the main disadvantages [65].

3.1.2 Osmopriming

Osmopriming is a priming method that uses osmotic solutions and activates metabolic processes before germination by providing lower water potential to the seeds. This practice takes advantage of the slow absorption of water through osmopriming agents. Osmopriming agents include sodium chloride, potassium chloride, potassium

Priming agent	Agent dose	Plant	Stress factor	Effect	Ref.
JA or SA	0.1 mM 0.5 mM	*Arabidopsis thaliana*	*Pseudomonas syringae* (PstDC3000)	General DNA hypomethylation changes. Enrichment of H3K9ac and H3K27me3 in WRKY6 and WRKY53 genes	[57]
2,6-dichloroisonicotinic acid (INA)	100 μM	*Phaseolus vulgaris*	*Pseudomonas syringae* pv. *phaseolicola* (NPS3121)	INA-primed common bean plants and its stress-free offsprings exhibited enrichment of H3K4me3 and H3K36me3 (transgenerational immune memory)	[58]
β-Aminobutyric acid (BABA)	5 mM	*Solanum lycopersicum*	*Botrytis cinerea*	Primed tomato leaves close to 50% reduction in the frequency of mCHH in the genome, and while the net reduction in CG and CHG methylation reflected a balance of both hyper- and hypomethylation, almost all CHH DMRs (differentially methylated regions) were hypomethylated.	[59]
β-Aminobutyric acid (BABA)	By spraying potato leaves with 5 mM of BABA (3 ml per plant)	*Solanum tuberosum*	*Phytophthora infestans*	High levels of the heritable H3K4me2 tag in BABA primed seeds (F0) and their vegetative, productive progeny (F1) provided evidence for the epigenetic marker for intergenerational memory in potato	[60]
NaCl	50 mM NaCl	*Arabidopsis thaliana*	Drought	Decrease in H3K27me3 at the chromatin level	[61]
NaCl	50 mM NaCl	*Arabidopsis thaliana*	Salinity stress	Increased expression of the Na-transporter HKT1 gene	[61]
NaCl	% 0.3 NaCl	*Glycine max*	Salinity stress	H3K4me2, H3K4me3, and H3K9ac were dramatically induced	[62]
Heavy metals	Cu^{2+} (1000 μM $CuSO_4$), Cd^{2+} (1000 μM $CdCl_2$), Cr^{3+} (1000 μM $CrCl_3$) or Hg^{2+} (50 μM $HgCl_2$)	*Oryza sativa*	Heavy metal	Changes associated with DNA methylation state of Tos17 retrotransposon	[63]

Table 2.
Epigenetic effects of priming agent on plant species.

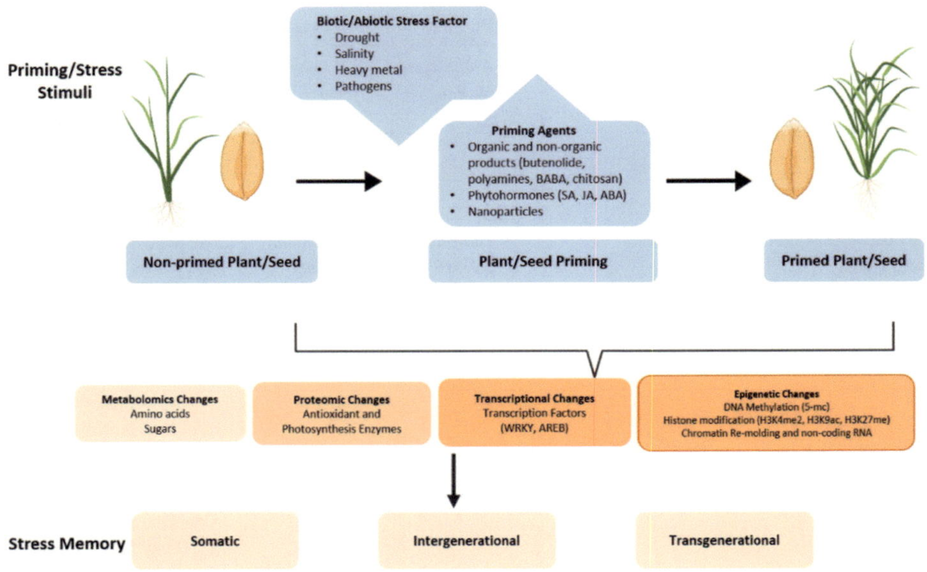

Figure 1.
Schematic representation of establishing stress memory through priming applications.

nitrate, calcium chloride, magnesium sulfate, mannitol, sorbitol, PEG (polyethylene glycol), and glycerol (**Table 3**). PEG is commonly used as osmopriming agent due to its large molecular size and nontoxic structure [92, 93]. Osmopriming is a more rapidly applicable technique compared to hydropriming. It provides farmers with a very appealing alternative to develop crop establishment and yield. Moreover, it has significantly lower cost than conventional water saving strategies [92, 94].

3.1.3 Chemical priming

Numerous natural or synthetic chemicals such as paclobutrazol, butenolide, choline, paclobutrazol, and chitosan are preferred as chemical preparation components to ensure that physiological and biological processes are not disrupted in plants under stress conditions (**Table 3**). Treatment with butenolide increases the germination and development rate in *Capsicum annum* L. and *Salvia* L. [64]. Selenium has a positive effect on crop growth and stress tolerance at low concentrations. Se priming of bitter gourd seeds has been reported to enhance antioxidant capacity and germination ability at low temperatures. Although chemical priming agents can be effective at very low concentrations, their use at high concentrations is considered harmful to the plant and the environment. For instance, NO and H_2S have inhibitory effects on the mitochondrial electron transport chain at high concentrations [95].

3.1.4 Biological priming

Microorganism-mediated biopriming (Plant Growth-Promoting Rhizobacteria (PGPRs), biocontrol agents, and fungicides) improves germination, seedling vigor, and biotic/abiotic stress tolerance and synchronizes crop stand. Microorganisms that are beneficial to plants such as *Trichoderma, Pseudomonas, Azotobacter*, *Azospirillum*, and *Agrobacterium* are used as biopriming agents (**Table 3**). Biopriming with

Seed priming method	Priming agent	Plant	Priming effect	Priming agent concentration/ duration	Ref.
Hydropriming	Water	*Zea mays*	Drought tolerance	36 h	[70]
Osmopriming	PEG	*Spinacia oleracea*	Osmopriming improved stress tolerance of germinating seeds/ increasing the germination potential of primed seeds	PEG 8000 (1, 2, 4 and 8 days/−0.6 MPa)	[71]
	KNO_3	*Helianthus annuus*	Increased salt tolerance of sunflower seeds by promoting K^+ and Ca^{2+} accumulation and inducing osmoregulation by the accumulation of proline	24 h KNO_3 (−1.0 MPa)	[72]
	$CaCl_2$	*Triticum aestivum*	Improved leaf area index, leaf area duration, and crop growth rate	−1.25 MPa	[73]
	Melatonin	*Brassica napus*	Improved seed germination under the drought stress, higher antioxidant activities	500 mM 6 h	[74]
Chemical priming	Paclobutrazol	*Solanum lycopersicum*	Drought tolerance	50 mg/L^{-1}	[75]
	Chitosan	*Zea mays*	Low temperature tolerance	0.50% (w/v)	[76]
	Putrecine (Put)	*Nicotiana tabacum*	Chilling tolerance	0.01 mM and 0.1 mM put for 24 h and 48 h	[77]
	Polyamines (putrecine, spermidine, spermine)	*Oryza sativa*	Drought tolerance	10 μM	[78]
	Spermidine (Spd)	*Oryza sativa*	Salt tolerance	1 mM-soaked seeds for 14 days	[79]
	NaHS and $CaCl_2$	*Cucurbita pepo*	Adaptation to Ni stress by increasing the AsA-GSH cycle, redox homeostasis	NaHS (100 μM) 24 h $CaCl_2$ (15 mM) 24 h	[80]
	Choline	*Triticum aestivum*	Salt tolerance	5 mM (24 h)	[81]
	KH_2PO_4	*Triticosecale Witm.*	Increased germination percentage under salt stress	0.5% (w/v)	[82]
Hormonal priming	Gibberellic acid	*Brassica napus*	Drought tolerance	300 mg/L 8 h	[83]
	Gibberellic acid, Salicylic acid	*Triticum aestivum*	Salicylic acid priming greatly enhanced the drought stress tolerance	10^{-4} M	[84]
	Salicylic acid	*Triticum aestivum*	Shortened the germination period and improved the germination rate	50, 75, 100 ppm/12 h	[85]

Seed priming method	Priming agent	Plant	Priming effect	Priming agent concentration/ duration	Ref.
Biopriming	Inoculation with *Rhizobium* and *Pseudomonas*	*Zea mays*	Salt tolerance	Cell density of 106 cells/ml	[86]
	Trichoderma harzianum	*Triticum aestivum*	Improving germination percentage and reducing reduction percentage of germination during salinity stress	10 g/kg of seeds	[87]
	Trichoderma harzianum	*Triticum aestivum*	Decrease in proline, MDA, and hydrogen peroxide	—	[88]
Nutripriming	Zinc ($ZnSO_4$)	*Triticum durum*	Higher antioxidant potential (SOD activity), better germination and development of seedlings	9, 20 and 50 mg Zn kg^1	[89]
		Triticum aestivum	Higher dissipation of excess energy, higher leaf succulence values	0.4% Zn 8 h	[90]
	$Mg(NO3)_2$ and $ZnSO_4$	*Tritucum aestivum*	Higher yield and attributes parameters (spike length, spike weight, seed count)	—	[91]

Table 3.
Seed priming-induced stress tolerance in plants.

Trichoderma lixii in maize (*Zea mays*) and *Trichoderma harzianum* in wheat (*Triticum aestivum* L.) resulted in photosynthetic activity-mediated salt stress tolerance [96]. Biopriming *Cicer arietinum* L. with *Rhizobacteria* under salt stress conditions improves seed germination percentage and increases root and shoot length [97].

3.1.5 Hormonal priming

Hormonal priming involves the application of solutions containing plant growth regulators to the seeds. Various regulators such as abscisic acid (ABA), SA (salicylic acid), gibberellic acid (GA3), ascorbate, kinetin, auxin, cytokinin, gibberellin, ethylene, and polyamines are used for priming (**Table 3**). Hormonal priming of *Glycine max* with benzyl adenine (4.87 mg L^{-1}), a cytokinin, was shown to increase plant growth, root biomass, flowering, and fruiting, significantly. These effects also suggest that benzyl adenine may play an important role in improving drought tolerance in *Glycine max* [98]. In another application, increased chlorophyll content and drought tolerances were achieved in *Zea mays* plant seeds primed with gibberellin [93].

3.1.6 Nutripriming

Nutripriming is a technique involving the application of solutions containing magnesium, zinc, and boron to the seeds. The mineral-nutrient status of plants plays a critical role in enhancing plant resistance to environmental stressors. Nutripriming with Zn in *Cicer arietinum* L. seed was found to play a role in the development of drought tolerance and improvement of yield characteristics [99]. Nutripriming of *Agropyron elongatum* with ascorbic acid under salt stress was found to increase the germination capacity of the plant [100].

3.2 Plant priming

In addition to numerous applications of priming to the seeds, plants can also get primed at the seedling stage. Soil irrigation and foliar spraying are the most widely used methods during priming application at seedling stage [64, 65]. Putrescine and spermine were applied to the wheat plants as chemical priming agents by spraying at the concentration of 0.1 mM. The mixture of putrescine and spermine not only improved *Tritucum aestivum* L. growth under drought stress but also increased RUBISCO levels, which enhances photosynthesis. Leaf spraying was found to be more effective than seed priming. It is suggested that the effectiveness of foliar spraying may be attributed to the improvement of water status in epidermal and underlying cell layers through direct contact between polyamines and the leaf surface [101]. Moreover, exogenous H_2O_2 priming application can change the response of *Pistacia vera* seedlings against NaCl stress after spraying 1, 5, and 10 mM H_2O_2 to 40-day-old seedlings. 0, 120, and 240 mM NaCl solutions were applied to the seedlings after 24 hours of application. A significant decrease in H_2O_2 and MDA contents were detected in the leaves primed with H_2O_2 after 7 days. H_2O_2-primed plants had higher enzymatic activity compared to non-primed plants. H_2O_2 can reduce lipid peroxidation (MDA content) and oxidative damage by increasing the activities of antioxidant enzymes (such as APX and CAT) or nonenzymatic antioxidant compounds (such as GSH, ASA, and CAR) under salt stress [102]. Abscisic acid, which is also used as a hormonal priming agent, increases the growth and survival rate of plants, when applied to 14-day-old rice (*Oryza sativa*) seedlings by root drenching (0 μM, 10 μM, and 50 μM). It regulates alkaline stress tolerance positively [103].

3.3 Priming and epigenetic

The ability of plants to recall previous stress after priming with biotic and abiotic stressors has been linked to epigenetic mechanisms. The accumulation of signaling molecules, DNA methylation, and histone modifications (HM) can be listed. For instance, histone deacetylases (HDACs) are involved in cellular processes such as chromatin structure and gene expression in the formation of stress responses to abiotic stresses by deacetylating histones. On the other hand, 5-methylcytosine (5mC), as another major epigenetic signal, is controlled by various DNA methyltransferases and demethylases. RNA-directed DNA methylation (RdDM) is an epigenetic process that mediates DNA methylation at specific DNA sequences of noncoding RNA. Together with DNA demethylation, RdDM plays a role in response to abiotic and biotic stresses by regulating the activity of transposable elements (TEs) and gene expression in both plant development and stress responses [104]. Recent studies have shown that epigenetic features such as histone acetylation and methylation can be altered by priming to alter chromatin state. This would alter transcriptional responses to a re-emerging stressor (**Table 2**). It has been suggested that stress exposure may modulate transcriptional responses [62]. High temperature stress has been associated with epigenetic marks such as methylation [5-methylcytosine (5mC), N6-methyladenine (6 mA)], siRNAs-controlled RdDMs, and histone modifications.

DNA methylation mediated by miRNAs is involved in salinity and drought stresses. Small RNAs, especially miRNAs, also contribute to the intergenerational inheritance of heat stress in *Brassica rapa* and *Tritucum durum* [104]. The priming agent BABA has been reported to induce DNA cytosine hypomethylation in *Solanum lycopersicum* [59]. The epigenetic marker H3K4me2 (histone H3 dimethylated at lysine 4) has been reported to contribute to the intergenerational transmission of *Solanum tuberosum* immune memory [60].

4. Concluding remarks and prospects

In this chapter, we focused on seed or plant priming applications and their future in agriculture to increase tolerance to abiotic and biotic stress conditions in important crop plants and to describe the mechanisms of triggering, protecting, and removing the stress memory in plants. We may suggest that combining priming and stress memory studies will be much more advantageous in elucidating the mechanisms involved in improving the stress response of plants. Seed priming methods can be used to increase plant performance under stressful conditions. We also should underline the fact that further studies and insight are required in many areas such as epigenetics of plant growth, development, and stress response in order to use these methods with maximum efficiency in the field.

The discovery of the cognitive abilities of plants to form and retain memory has offered scientists and plant breeders exciting new opportunities. Primary stress exposure or stress priming, as a nongenetic approach, has been considered as one of the novel approaches to improve crop tolerance and resistance, regulating the genetic and epigenetic processes that govern the stress memory of plants. Plants can sense the conditions in their environment and integrate over time and store environmental cues. Plants become more responsive to the stresses that are repeated over and over in subsequent generations in conjunction with changes coordinated with stress memory at the cellular levels. Acquired tolerance refers to memory that occurs due to stress.

Priming has some potential to produce more tolerant and productive crops in the future. However, its relation to molecular mechanisms and stress memory has not yet been fully understood. Like the genome, transcriptome, and proteome of plants, their epigenome also contributes to stress responses. As a result, primed plants respond more rapidly and effectively to biotic or abiotic stress conditions. Stress priming, also known as training or conditioning, can increase short- or long-term stress memory of plants. This allows plants to get more resistant to more stresses in current and even subsequent generations. It has been shown in many studies that mitotic stress memory, which affects the somatic memory of a generation, and meiotic stress memory, which affects the memory of future generations, can play an important role in the development of plant responses to stresses. Somatic memory is typically transient and is activated when exposed to acute stress but can be reactivated typically within a limited time frame, that is, a few hours or days. Various priming strategies have been applied, such as accumulation of various cellular compounds, phosphorylation of mitogen-activated protein kinases (MAPKs), and modification of regulatory proteins through epigenetic mechanisms [53]. In plants, somatic memory is achieved through a variety of mechanisms such as chromatin remodeling, alternative transcript splicing, metabolite accumulation, and autophagy [105].

Besides advantages such as ease of application and cost-effectiveness, there are also disadvantages such as the risks of causing an increase in the amount of chemicals in agricultural areas and the difficulties in scale-up applications in large fields. Moreover, the necessity of dose optimization and application time depending on the variety is also a major obstacles. Priming synchronizes seed germination and facilitates seedling formation and plant growth under stressful conditions; hence, it supports yield increase. However, there are important gaps in the development of tolerance and resistance by stress priming and stress memory in plants. There are many aspects that need to be clarified in the future. Intergenerational memory, which is passed directly from the first generation to the next, can be reset in the second generation. As stress memory is passed on to offspring, it can be reset or cause negative impact in plants due to phenotypic or physiological stress plasticity. For this reason, in-field validation is essential, and more research is required to fully understand the whole stress memory regulation. Understanding the molecular basis of stress priming will contribute to identifying potential stress response options including cellular targets and signaling networks.

Seed priming is a pre-germination process and creates mild stress in the early stages of germination through various chemical, physical, and biological agents. To maximize the cost/benefit ratio for farmers, it is crucial to identify and cultivate the local varieties that best respond to seed priming. Contrary to the conditions in the laboratory, plants encounter many stresses in various combinations at the same time in their natural environment. Therefore, further studies are required to understand how epigenetic changes and responsible pathways are adjusted against different abiotic or biotic stresses, priming doses, and durations. Determination and optimization of the plant development stages to initiate priming can also enhance effectiveness and efficiency. Research on the regulation of epigenetic stress responses in crops needs to be accelerated. Data on DNA methylation should be evaluated with other epigenetic changes in establishing stress tolerance. Thus, the roles of epigenetic changes induced by stress in regulation of the expression of crucial genes involves signal perception or transduction. It should be considered that the development of plants under optimal conditions is somewhat compromised by seed priming to alter the somatic or transgenerational memory of plants. Crop productivity may be adversely affected as the quality of seeds with the best germination rate is reduced due to applied prestress. Their seedling growth may also potentially get interrupted. From this point of view, the balance between plant growth and stress tolerance should be carefully evaluated. With a variety of omics approaches (proteomics, metabolomics, transcriptomics, and epigenomics) and genome-wide association studies (GWASs) using high-throughput methods, it is possible to learn the molecular basis of stress memory of plants and generate data to correlate molecular function with agronomic performance.

When abiotic and biotic stresses occur, it is vital to recall molecular experiences and utilize recorded information to adapt to new conditions. Priming and stress memory are novel focal points for crop management and sustainable agriculture. While plants with rearranged stress memory offer a new possibilities for creating highly productive agricultural practices with better stress responses, it is an undeniable fact that much research is required to develop this strategy and apply it effectively in field.

Author details

Tamer Gümüş, Tuğçe Aydın, Burcu Gündüz, Sinan Meriç, Alp Ayan* and Çimen Atak
Department of Molecular Biology and Genetic, Faculty of Science and Letters,
Istanbul Kultur University, Istanbul, Turkey

*Address all correspondence to: a.ayan@iku.edu.tr

References

[1] Ogruc Ildiz G, Celik O, Atak C, et al. Raman spectroscopic and chemometric investigation of lipid–protein ratio contents of soybean mutants. Applied Spectroscopy. 2020;**74**:34-41. DOI: 10.1177/00037028198599

[2] Çelik Ö, Ayan A, Meriç S, et al. Comparison of tolerance related proteomic profiles of two drought tolerant tomato mutants improved by gamma radiation. Journal of Biotechnology. 2021;**330**:35-44. DOI: 10.1016/j.jbiotec.2021.02.012

[3] Ayan A, Meriç S, Gümüş T, et al. In: Ohyama T, Takahashi Y, Ohtake N, Sato T, Tanabata S, editors. Current Strategies and Future of Mutation Breeding in Soybean Improvement, Soybean—Recent Advances in Research and Applications. London, UK: IntechOpen; 2022. DOI: 10.5772/intechopen.104796

[4] Meriç S, Ayan A, Atak Ç, et al. Profile-based proteomic investigation of unintended effects on transgenic and gamma radiation induced mutant soybean plants. Genetic Resources and Crop Evolution. 2023;**70**(2077-2095):1-19. DOI: 10.1007/s10722-023-01560-5

[5] Meriç S, Ayan A, Atak Ç. In: Fahad S, Saud S, Chen Y, Wu C, Wang D, editors. Molecular abiotic stress tolerans strategies: From genetic engineering to genome editing era, Abiotic Stress Plants. London, UK: IntechOpen; 2020. DOI: 10.5772/intechopen.94505

[6] Meriç S, Gümüş T, Ayan A. Plant-based vaccines: The future of preventive healthcare? In: Ghimire K, editor. Botany—Recent Advances and Applications. London, UK: IntechOpen; 2021. DOI: 10.5772/intechopen.97861

[7] Ayan A, Meriç S, Gümüş T, et al. Next generation of transgenic plants: From farming to pharming. In: Sithole Niang I, editor. Genetically Modified Plants and Beyond. London, UK, London, UK: IntechOpen; 2021. DOI: 10.5772/intechopen.102004

[8] Ayan A, Meriç S, Gümüş T, et al. Transgenic plants in heat stress adaptation: Present achievements and prospects. In: Oliveira M, Fernandes-Silva A, editors. Abiotic Stress in Plants—Adaptations to Climate Change. London, UK: IntechOpen; 2023. DOI: 10.5772/intechopen.111791

[9] Meriç S, Çakır Ö, Turgut-Kara N, et al. Detection of genetically modified maize and soybean in feed samples. Genetics and Molecular Research: GMR. 2014;**13**:1160-1168. DOI: 10.4238/2014. February.25.2

[10] Çakır Ö, Meriç S, Arı Ş. GMO analysis methods for food: From today to tomorrow. Food Safety. In: Si UG, Giuseppe C, editors. Food Safety: Innovative Analytical Tools for Safety Assessment. New York: WILEY-Scrivener Publishing; 2017. pp. 123-179. DOI: 10.1002/9781119160588

[11] Aranega-Bo P, de la O Leyva M, Finiti I, et al. Priming of plant resistance by natural compounds. Frontiers in Plant Science. 2014;5:488. DOI: 10.3389/fpls.2014.00488

[12] Harris CJ, Amtmann A, Ton J. Epigenetic processes in plant stress priming: Open questions and new approaches. Current Opinion in Plant Biology. 2023;**75**:102432. DOI: 10.1016/j.pbi.2023.102432

[13] Nosaka M, Itoh JI, Nagato Y, et al. Role of transposon-derived small RNAs

in the interplay between genomes and parasitic DNA in Rice. PLoS Genetics. 2012;**8**:e1002953. DOI: 10.1371/journal.pgen.1002953

[14] Chang YN, Zhu C, Jiang J, et al. Epigenetic regulation in plant abiotic stress responses. Journal of Integrative Plant Biology. 2020;**62**:563-580. DOI: 10.1111/jipb.12901

[15] Lucibelli F, Valoroso MC, Aceto S. Plant DNA methylation: An epigenetic mark in development, environmental interactions, and evolution. International Journal of Molecular Sciences. 2022;**23**:8299. DOI: 10.3390/ijms23158299

[16] Li M, Da FY. Histone variants: The artists of eukaryotic chromatin. Science China Life Sciences. 2015;**58**:232-239. DOI: 10.1007/s11427-015-4817-4

[17] Struhl K, Segal E. Determinants of nucleosome positioning. Nature Structural & Molecular Biology. 2013;**20**:267-273. DOI: 10.1038/nsmb.2506

[18] Foroozani M, Holder DH, Deal RB. Histone variants in the specialization of plant chromatin. Annual Review of Plant Biology. 2022;**73**:149-172. DOI: 10.1146/annurev-arplant-070221-050044

[19] Zhao T, Zhan Z, Jiang D. Histone modifications and their regulatory roles in plant development and environmental memory. Journal of Genetics and Genomics. 2019;**46**:467-476. DOI: 10.1016/j.jgg.2019.09.005

[20] Zhang C, Du X, Tang K, et al. Arabidopsis AGDP1 links H3K9me2 to DNA methylation in heterochromatin. Nature Communications. 2018;**9**:1-14. DOI: 10.1038/s41467-018-06965-w

[21] Brewis HT, Wang AY, Gaub A, et al. What makes a histone variant a variant: Changing H2A to become H2A.Z. PLoS Genetics. 2021;**17**:e1009950. DOI: 10.1371/journal.pgen.1009950

[22] Liu H, Able AJ, Able JA. Small RNAs and their targets are associated with the transgenerational effects of water-deficit stress in durum wheat. Scientific Reports. 2021;**11**:1-17. DOI: 10.1038/s41598-021-83074-7

[23] Kambona CM, Koua PA, Léon J, et al. Stress memory and its regulation in plants experiencing recurrent drought conditions. Theoretical and Applied Genetics. 2023;**136**:1-21. DOI: 10.1007/s00122-023-04313-1

[24] Bhadouriya SL, Mehrotra S, Basantani MK, et al. Role of chromatin architecture in plant stress responses: An update. Frontiers in Plant Science. 2021;**11**:603380. DOI: 10.3389/fpls.2020.603380

[25] Review M, Song ZT, Liu JX, et al. Chromatin remodeling factors regulate environmental stress responses in plants. Journal of Integrative Plant Biology. 2021;**63**:438-450. DOI: 10.1111/jipb.13064

[26] Kim JH. Multifaceted chromatin structure and transcription changes in plant stress response. International Journal of Molecular Sciences. 2021;**22**:2013. DOI: 10.3390/ijms22042013

[27] Liu N, Staswick PE, Avramova Z. Memory responses of jasmonic acid-associated Arabidopsis genes to a repeated dehydration stress. Plant, Cell & Environment. 2016;**39**:2515-2529. DOI: 10.1111/pce.12806

[28] Ding Y, Fromm M, Avramova Z. Multiple exposures to drought 'train' transcriptional responses in Arabidopsis.

Nature Communications. 2012;**3**:1-9. DOI: 10.1038/ncomms1732

[29] Crisp PA, Ganguly D, Eichten SR, et al. Reconsidering plant memory: Intersections between stress recovery, RNA turnover, and epigenetics. Science Advances. 2016;**2**:e1501340. DOI: 10.1126/sciadv.150134

[30] de Freitas A, Guedes F, Menezes-Silva PE, DaMatta FM, et al. Using transcriptomics to assess plant stress memory. Theoretical and experimental. Plant Physiology. 2019;**31**:47-58. DOI: 10.1007/s40626-018-0135-0

[31] Métraux JP, Signer H, Ryals J, et al. Increase in salicylic acid at the onset of systemic acquired resistance in cucumber. Science. 1990;**250**:1004-1006. DOI: 10.1126/science.250.4983.10

[32] Jacques C, Salon C, Barnard RL, et al. Drought stress memory at the plant cycle level: A review. Plants. 2021;**10**:1873. DOI: 10.3390/plants10091873

[33] Lämke J, Bäurle I. Epigenetic and chromatin-based mechanisms in environmental stress adaptation and stress memory in plants. Genome Biology. 2017;**18**:1-11. DOI: 10.1186/s13059-017-1263-6

[34] Sharma M, Kumar P, Verma V, et al. Understanding plant stress memory response for abiotic stress resilience: Molecular insights and prospects. Plant Physiology and Biochemistry. 2022;**179**:10-24. DOI: 10.1016/j.plaphy.2022.03.004

[35] Zuo DD, Ahammed GJ, Guo DL. Plant transcriptional memory and associated mechanism of abiotic stress tolerance. Plant Physiology and Biochemistry. 2023;**201**:107917. DOI: 10.1016/j.plaphy.2023.107917

[36] Oberkofler V, Pratx L, Bäurle I. Epigenetic regulation of abiotic stress memory: Maintaining the good things while they last. Current Opinion in Plant Biology. 2021;**61**:102007. DOI: 10.1016/j.pbi.2021.102007

[37] Jaskiewicz M, Conrath U, Peterhälnsel C. Chromatin modification acts as a memory for systemic acquired resistance in the plant stress response. EMBO Reports. 2011;**12**:50-55. DOI: 10.1038/embor.2010.186

[38] Brzezinka K, Altmann S, Czesnick H, et al. Arabidopsis FORGETTER1 mediates stress-induced chromatin memory through nucleosome remodeling. eLife. 2016;**5**:e17061. DOI: 10.7554/eLife.17061

[39] Wibowo A, Becker C, Marconi G, et al. Hyperosmotic stress memory in arabidopsis is mediated by distinct epigenetically labile sites in the genome and is restricted in the male germline by dna glycosylase activity. eLife. 2016;**5**:e13546. DOI: 10.7554/eLife.13546

[40] Kuźnicki D, Meller B, Arasimowicz-Jelonek M, et al. BABA-induced DNA methylome adjustment to intergenerational defense priming in potato to Phytophthora infestans. Frontiers in Plant Science. 2019;**10**:445885. DOI: 10.3389/fpls.2019.00650

[41] Whittle CA, Otto SP, Johnston MO, et al. Adaptive epigenetic memory of ancestral temperature regime in *Arabidopsis thaliana*. Botany. 2009;**87**:650-657. DOI: 10.1139/B09-030

[42] Nosalewicz A, Siecińska J, Śmiech M, et al. Transgenerational effects of temporal drought stress on spring barley morphology and functioning. Environmental and Experimental

Botany. 2016;**131**:120-127. DOI: 10.1016/j.envexpbot.2016.07.006

[43] Hatzig SV, Nuppenau JN, Snowdon RJ, et al. Drought stress has transgenerational effects on seeds and seedlings in winter oilseed rape (*Brassica napus* L.). BMC Plant Biology. 2018;**18**:1-13. DOI: 10.1186/s12870-018-1531-y

[44] Bonasio R, Tu S, Reinberg D. Molecular signals of epigenetic states. Science. 2010;**330**:612-616. DOI: 10.1126/science.1191078

[45] Ali M, Sugimoto K, Ramadan A, et al. Memory of plant communications for priming anti-herbivore responses. Scientific Reports. 2013;**3**:1872. DOI: 10.1038/srep01872

[46] Berry S, Dean C. Environmental perception and epigenetic memory: Mechanistic insight through FLC. The Plant Journal. 2015;**83**:133-148. DOI: 10.1111/tpj.12869

[47] Hilker M, Schmülling T. Stress priming, memory, and signalling in plants. Plant, Cell & Environment. 2019;**42**:753-761. DOI: 10.1111/pce.13526

[48] Avramova Z. Transcriptional 'memory' of a stress: Transient chromatin and memory (epigenetic) marks at stress-response genes. The Plant Journal. 2015;**83**:149-159. DOI: 10.1111/tpj.12832

[49] Hauser MT, Aufsatz W, Jonak C, et al. Transgenerational epigenetic inheritance in plants. Biochimica et Biophysica Acta (BBA)-Gene Regulatory Mechanisms. 2011;**1809**:459, 10.1016/j.bbagrm.2011.03.007

[50] Sedaghatmehr M, Thirumalaikumar VP, Kamranfar I, et al. A regulatory role of autophagy for resetting the memory of heat stress in plants. Plant, Cell & Environment. 2019;**42**:1054-1064. DOI: 10.1111/pce.13426

[51] Chen H, Dong J, Wang T. Autophagy in plant abiotic stress management. International Journal of Molecular Sciences. 2021;**22**:22. DOI: 10.3390/ijms22084075

[52] Calarco JP, Borges F, Donoghue MTA, et al. Reprogramming of DNA methylation in pollen guides epigenetic inheritance via small RNA. Cell. 2012;**151**:194-205. DOI: 10.1016/j.cell.2012.09.001

[53] Turgut-Kara N, Arikan B, Celik H. Epigenetic memory and priming in plants. Genetica. 2020;**148**:47-54. DOI: 10.1007/s10709-020-00093-4

[54] Singh RR, Pajar JA, Audenaert K, et al. Induced resistance by ascorbate oxidation involves potentiating of the phenylpropanoid pathway and improved rice tolerance to parasitic nematodes. Frontiers in Plant Science. 2021;**12**:713870. DOI: 10.3389/fpls.2021.713870

[55] Yang Z, Zhi P, Chang C. Priming seeds for the future: Plant immune memory and application in crop protection. Frontiers in Plant Science. 2022;**13**:961840. DOI: 10.3389/fpls.2022.961840

[56] Liu H, Able AJ, Able JA. Priming crops for the future: Rewiring stress memory. Trends in Plant Science. 2022;**27**:699-716. DOI: 10.1016/j.tplants.2021.11.015

[57] Luna E, Bruce TJA, Roberts MR, et al. Next-generation systemic acquired resistance. Plant Physiology. 2012;**158**:844-853. DOI: 10.1104/pp.111.187468

[58] Martínez-Aguilar K, Hernández-Chávez JL, Alvarez-Venegas R. Priming of seeds with INA and its transgenerational effect in common bean (*Phaseolus vulgaris* L.) plants. Plant Science. 2021;**305**:110834. DOI: 10.1016/j.plantsci.2021.110834

[59] Catoni M, Alvarez-Venegas R, Worrall D, et al. Long-lasting defence priming by β-aminobutyric acid in tomato is marked by genome-wide changes in DNA methylation. Frontiers Plant Science. 2022;**13**:836326. DOI: 10.3389/fpls.2022.836326

[60] Meller B, Kuźnicki D, Arasimowicz-Jelonek M, et al. Baba-primed histone modifications in potato for intergenerational resistance to phytophthora infestans. Front Plant Science. 2018;**9**:1228. DOI: 10.3389/fpls.2018.01228

[61] Sani E, Herzyk P, Perrella G, et al. Hyperosmotic priming of Arabidopsis seedlings establishes a long-term somatic memory accompanied by specific changes of the epigenome. Genome Biology. 2013;**14**:1-24. DOI: 10.1186/gb-2013-14-6-r59

[62] Yung WS, Wang Q, Huang M, et al. Priming-induced alterations in histone modifications modulate transcriptional responses in soybean under salt stress. Plant Journal. 2022;**109**:1575-1590. DOI: 10.1111/tpj.15652

[63] Cong W, Miao Y, Xu L, et al. Transgenerational memory of gene expression changes induced by heavy metal stress in rice (*Oryza sativa* L.). BMC Plant Biology. 2019;**19**:1-14. DOI: 10.1186/s12870-019-1887-7

[64] Jisha KC, Vijayakumari K, Puthur JT. Seed priming for abiotic stress tolerance: An overview. Acta Physiologiae Plantarum. 2012;**35**:1381-1396. DOI: 10.1007/s11738-012-1186-5

[65] Marthandan V, Geetha R, Kumutha K, et al. Seed priming: A feasible strategy to enhance drought tolerance in crop plants. International Journal of Molecular Sciences. 2020;**21**:8258. DOI: 10.3390/ijms21218258

[66] Devika OS, Singh S, Sarkar D, et al. Seed priming: A potential supplement in integrated resource management under fragile intensive ecosystems. Frontiers in Sustainable Food Systems. 2021;**5**:654001. DOI: 10.3389/fsufs.2021.654001

[67] Pawar VA, Laware SL. Seed priming a critical review. International Journal of Scientific Research in Biological Sciences. 2018;**5**:94-101

[68] Villagómez-Aranda AL, Feregrino-Pérez AA, García-Ortega LF, et al. Activating stress memory: Eustressors as potential tools for plant breeding. Plant Cell Reports. 2022;**41**:1481-1498. DOI: 10.1007/s00299-022-02858-x

[69] Lal SK, Kumar S, Sheri V, et al. Seed priming: An emerging technology to impart abiotic stress tolerance in crop plants. In: Rakshit A, Singh H, editors. Advances in Seed Priming. Springer; 2018. pp. 41-50. DOI: 10.1007/978-981-13-0032-5_3

[70] Janmohammadi M, Dezfuli P, Physiol FS-GAP, et al. Seed invigoration techniques to improve germination and early growth of inbred line of maize under salinity and drought stress. General and Applied Plant Physiology. 2008;**34**:215-226

[71] Chen K, Arora R. Dynamics of the antioxidant system during seed osmopriming, post-priming

germination, and seedling establishment in spinach (*Spinacia oleracea*). Plant Science. 2011;**180**:212-220. DOI: 10.1016/j.plantsci.2010.08.007

[72] Bajehbaj AA. The effects of NaCl priming on salt tolerance in sunflower germination and seedling grown under salinity conditions. African Journal of Biotechnology. 2010;**9**:1764-1770. DOI: 10.5897/AJB10.1019

[73] Hussain M, Farooq M, Sattar A, et al. Mitigating the adverse effects of drought stress through seed priming and seed quality on wheat (*Triticum aestivum* L.) productivity. Pakistan Journal of Agricultural Sciences. 2023;**55**:313-319. DOI: 10.21162/PAKJAS/185833

[74] Khan MN, Zhang J, Luo T, et al. Seed priming with melatonin coping drought stress in rapeseed by regulating reactive oxygen species detoxification: Antioxidant defense system, osmotic adjustment, stomatal traits and chloroplast ultrastructure perseveration. Industrial Crops and Products. 2019;**140**:111597. DOI: 10.1016/j.indcrop.2019.111597

[75] Souza-Machado V, Pitblado R, Ali A, et al. Paclobutrazol in tomato (*Lycopersicon esculentum*) for improved tolerance to early transplanting and earlier harvest maturity. Acta Horticulture. 1999;**487**:139-143. DOI: 10.17660/ActaHortic.1999.487.17

[76] Guan YJ, Hu J, Wang XJ, et al. Seed priming with chitosan improves maize germination and seedling growth in relation to physiological changes under low temperature stress. Journal of Zhejiang University Science B. 2009;**10**:427-433. DOI: 10.1631/jzus.B0820373

[77] Xu S, Hu J, Li Y, et al. Chilling tolerance in Nicotiana tabacum induced by seed priming with putrescine. Plant Growth Regulation. 2011;**63**:279-290. DOI: 10.1007/s10725-010-9528-z

[78] Farooq M, Wahid A, Lee DJ. Exogenously applied polyamines increase drought tolerance of rice by improving leaf water status, photosynthesis and membrane properties. Acta Physiologiae Plantarum. 2009;**31**:937-945. DOI: 10.1007/s11738-009-0307-2

[79] Chunthaburee S, Sanitchon J, Pattanagul W, et al. Alleviation of salt stress in seedlings of black glutinous rice by seed priming with spermidine and gibberellic acid. Notulae Botanicae Horti Agrobotanici Cluj-Napoca. 2014;**42**:405-413. DOI: 10.15835/nbha4229688

[80] Valivand M, Amooaghaie R, Ahadi A. Seed priming with H2S and Ca2+ trigger signal memory that induces cross-adaptation against nickel stress in zucchini seedlings. Plant Physiology and Biochemistry. 2019;**143**:286-298. DOI: 10.1016/j.plaphy.2019.09.016

[81] Salama K, Mansour M, Sci NH-AJBA, et al. Choline priming improves salt tolerance in wheat (*Triticum aestivum* L.). Australian Journal of Basic and Applied Sciences. 2011;**5**:126-132. DOI: 10.13140/2.1.4228.9606

[82] Yagmur M, Kaydan D. Alleviation of osmotic stress of water and salt in germination and seedling growth of triticale with seed priming treatments. African Journal of Biotechnology. 2008;**7**:2156-2162

[83] Li Z, Lu GY, Zhang XK, et al. Improving drought tolerance of germinating seeds by exogenous application of gibberellic acid (GA3) in rapeseed (*Brassica napus* L.). Seed Science and Technology. 2010;**38**:432-440. DOI: 10.15258/sst.2010.38.2.16

[84] Ulfat A, Majid S, Bot AH-PJ, et al. Hormonal seed priming improves wheat (*Triticum aestivum* L.) field performance under drought and non-stress conditions. Pakistan Journal of Botany. 2017;**49**:1239-1253

[85] Jatana BS, Ram H, Gupta N. Application of seed and foliar priming strategies to improve the growth and productivity of late sown wheat (*Triticum aestivum* L.). Cereal Research Communications. 2020;**48**:383-390. DOI: 10.1007/s42976-020-00036-x

[86] Bano A, Fatima M. Salt tolerance in *Zea mays* (L). Following inoculation with Rhizobium and Pseudomonas. Biology and Fertility of Soils. 2009;**45**:405-413. DOI: 10.1007/s00374-008-0344-9

[87] Rawat L, Singh Y, Shukla N, et al. Alleviation of the adverse effects of salinity stress in wheat (*Triticum aestivum* L.) by seed biopriming with salinity tolerant isolates of *Trichoderma harzianum*. Plant and Soil. 2011;**347**:387-400. DOI: 10.1007/s11104-011-0858-z

[88] Shukla N, Awasthi RP, Rawat L, et al. Seed biopriming with drought tolerant isolates of *Trichoderma harzianum* promote growth and drought tolerance in *Triticum aestivum*. Annals of Applied Biology. 2015;**166**:171-182. DOI: 10.1111/aab.12160

[89] Candan N, Cakmak I, Ozturk L. Zinc-biofortified seeds improved seedling growth under zinc deficiency and drought stress in durum wheat. Journal of Plant Nutrition and Soil Science. 2018;**181**:388-395. DOI: 10.1002/jpln.201800014

[90] Pavia I, Roque J, Rocha L, et al. Zinc priming and foliar application enhances photoprotection mechanisms in drought-stressed wheat plants during anthesis. Plant Physiology and Biochemistry. 2019;**140**:27-42. DOI: 10.1016/j.plaphy.2019.04.028

[91] Kumar Singhal R, Pradesh U, Vivek Kumar I, et al. Improving the yield and yield attributes in wheat crop using seed priming under drought stress. Journal of Pharmacognosy and Phytochemistry. 2019;**8**:214-220

[92] Bhanuprakash K, Yogeesha HS. Seed priming for abiotic stress tolerance: An overview. In: Rao N, Shivashankara K, Laxman R, editors. Abiotic Stress Physiology of Horticultural Crops. Springer; 2016. pp. 103-117. DOI: 10.1007/978-81-322-2725-0_6

[93] Rhaman MS, Imran S, Rauf F, et al. Seed priming with phytohormones: An effective approach for the mitigation of abiotic stress. Plants. 2021;**10**:37. DOI: 10.3390/plants10010037

[94] Farooq M, Hussain M, Habib M. Influence of seed priming techniques on grain yield and economic returns of bread wheat planted at different spacings. Crop and Pasture Science. 2020;**71**:725-738. DOI: 10.1071/cp20065

[95] Savvides A, Ali S, Tester M, et al. Chemical priming of plants against multiple abiotic stresses: Mission possible? Trends Plant Science. 2016;**21**:329-340. DOI: 10.1016/j.tplants.2015.11.003

[96] Chakraborti S, Bera K, Sadhukhan S, et al. Bio-priming of seeds: Plant stress management and its underlying cellular, biochemical and molecular mechanisms. Plant Stress. 2022;**3**:100052. DOI: 10.1016/j.stress.2021.100052

[97] Mishra M, Kumar Madan Mohan U, Prakash V, et al. Efficiency of plant growth promoting rhizobacteria for the enhancement of *Cicer arietinum* L. growth and germination under salinity.

Advances in Biological Research. 2010;**4**:92-96

[98] Mangena P. Effect of hormonal seed priming on germination, growth, yield and biomass allocation in soybean grown under induced drought stress. Indian Journal of Agricultural Research. 2020;**54**:592-598. DOI: 10.18805/IJARe.A-441

[99] Singh Shivay Y, Prasad R, Pal M. Genetic variability for zinc use efficiency in chickpea as influenced by zinc fertilization. International Journal of Bio-resource Stress Management. 2014;**5**:031-036. DOI: 10.5958/j.0976-4038.5.1.005

[100] Tavili A, Zare S, Enayati A. Hydropriming, ascorbic and salicylic acid influence on germination of *Agropyron elongatum* host. Seeds under salt stress. Research Journal of Seed Science. 2009;**2**:16-22. DOI: 10.3923/rjss.2009.16.22

[101] Hassan N, Ebeed H, Aljaarany A. Exogenous application of spermine and putrescine mitigate adversities of drought stress in wheat by protecting membranes and chloroplast ultra-structure. Physiology and Molecular Biology of Plants. 2020;**26**:233-245. DOI: 10.1007/s12298-019-00744-7

[102] Bagheri M, Gholami M, Baninasab B. Hydrogen peroxide-induced salt tolerance in relation to antioxidant systems in pistachio seedlings. Scientia Horticulturae. 2019;**243**:207-213. DOI: 10.1016/j.scienta.2018.08.026

[103] Wei LX, Lv BS, Wang MM, et al. Priming effect of abscisic acid on alkaline stress tolerance in rice (*Oryza sativa* L.) seedlings. Plant Physiology and Biochemistry. 2015;**90**:50-57. DOI: 10.1016/j.plaphy.2015.03.002

[104] Gallusci P, Agius DR, Moschou PN, et al. Deep inside the epigenetic memories of stressed plants. Trends Plant Science. 2023;**28**:142-153. DOI: 10.1016/j.tplants.2022.09.004

[105] Srivastava AK, Suresh Kumar J, Suprasanna P. Seed 'primeomics': Plants memorize their germination under stress. Biological Reviews. 2021;**96**:1723-1743. DOI: 10.1111/brv.12722

Chapter 5

Accelerated Approaches for Cabbage Improvement

Shipra Singh Parmar, Impa H. Ravindra and Ramesh Kumar

Abstract

Cabbage is widely recognized as a good source of dietary fiber, minerals, vitamins C and provitamin A carotenoids and some glucosinolates that may have a chemoprotective impact in humans. It is a highly cross-pollinated crop where heterosis in F_1 hybrid progeny has been exploited for development of hybrids. The self-incompatibility and male sterility systems are present in the crop, which facilitates easy and cheaper hybrid production. Different conventional and biotechnological approaches are being utilized for the improvement of cabbage. Modern breeding approaches such as marker-assisted breeding and transgenic approaches such as *Agrobacterium*-mediated gene transfer and through genome editing techniques, which offer a new opportunity for genetic improvement of the cabbage. The molecular markers represent a useful resource for enhancing selection efficiency via marker-assisted selection (MAS) in cabbage breeding.

Keywords: cabbage, breeding, genetic improvement, hybrid, modern breeding

1. Introduction

Cabbage, scientifically known as *Brassica oleracea* L. var. *capitata* L., holds significant prominence as a cole crop cultivated worldwide. It possesses a diploid chromosome count of 2n = 2x = 18. Contemporary cabbage varieties have been derived from wild, non-heading Brassica (*Brassica oleraceae* L. var. *capitata* L.) through processes such as mutation, human selection, and adaptation. The origin of this phenomenon is commonly attributed to the East Mediterranean and Asia Minor regions. Various varieties of cabbage exist, distinguished by their respective characteristics such as shape, size, color, and leaf structure [1]. Cabbage is cultivated primarily for its leafy heads, which are frequently utilized in various culinary preparations such as salads, cole slaw, boiled vegetables, pickled dishes, and dehydrated products. Additionally, cabbage leaves can be fermented under pressure to produce sauerkraut. Cabbage is known to contain amino acids, minerals, ß-carotene, and ascorbic acid [2]. The substance in question exhibits a high concentration of proteins, minerals, and antioxidants, which have been found to possess anti-carcinogenic properties [3, 4] and anti-obesity properties [5]. Cabbage has been found to possess a protective effect against bowel cancer, which can be attributed to the presence of the compound indole-3-carbinol [6]. India is the second largest producer of cabbage globally, following China. The global cultivation of the crop spans across an expansive area of 2414 thousand hectares, yielding a substantial production of 70862.50 thousand tonnes. This equates to a productivity rate of 29.3 tonnes per hectare [7]. In the specific context of India, the crop is cultivated on a smaller scale, covering an area of 397 thousand hectares and resulting in a production of 9207 thousand tones [8].

2. Origin and evolution

The cultivated cabbages that exist today are derived from the wild cabbage species known as *Brassica oleracea* L., commonly referred to as colewort or field cabbage. The species in question is indigenous to the coastal regions of Western Europe and the Western Mediterranean. It thrives in challenging environments, specifically on ledges of chalky cliffs and even on near-vertical rocky surfaces where other plant species are unable to establish themselves [9]. The Chinese cabbage, turnip, rutabaga, and oil-seed rape have been derived from various *Brassica* species [10]. The cultivation of cabbage was documented in Germany by the year 1150, and there is evidence to suggest that it may have been introduced to England as early as the 14th century. The cultivation of this crop has been practiced in China since ancient times. The initial introduction of cabbage into the United States, specifically Virginia, can be attributed to the English colonizers [11]. The introduction of cabbage into Canada occurred in 1541 through the efforts of Jacques Cartier. The introduction of cabbage into India by the Portuguese predates the introduction of cauliflower by the British. However, its popularity only emerged during the period of British colonial rule [12].

The wild cabbage, a leafy winter annual that is found along the coasts of the North Sea, the English Channel, and the northern Mediterranean Sea, is considered the most probable candidate for an ancestral form. It is thought to have its origins in southern Europe and spread to other places by humans [13].

Based on current knowledge, it is indicated that contemporary cultivated crops have originated from the wild *B. oleracea* rather than the wild Mediterranean species, despite the likelihood of initial selection of various crop varieties taking place in the Mediterranean region [14]. The distribution of wild *B. oleracea* in the Mediterranean region appears to be improbable based on current evidence. The prevailing view suggests that early cultivated variations of *B. oleracea* were introduced from the Atlantic coast to the Mediterranean, where subsequent selection processes led to the development of various crop types. Nevertheless, it is important to acknowledge that the matter remains open to differing viewpoints. For instance, [15] all cole crops originated from the Mediterranean region, primarily through mutation and introgression from wild species during evolution or by human selection.

3. Cytogenetics

Cabbage and its botanical relatives exhibit a diploid chromosomal count, denoted as $2n = 2x = 18$. Nevertheless, the examination of secondary chromosome associations could potentially imply that the value of x is either 5 or 6, thereby indicating that cabbage could be classified as a modified amphidiploids [16]. The cabbage plant possesses a somatic chromosome count of $2n = 18$, with its genome being denoted as c [17]. The size of cabbage chromosomes is relatively diminutive. The classification of the genome of nine has been conducted based on its size. The chromosomal composition consists of one extremely elongated chromosome, four chromosomes of considerable

Leaf color		
A	Basic anthocyanin development factor series, intensifiers	[22]
Arc	Colored lamina; color intensifier in red cabbage	[22]
B	Light red midrib alone; with A gives a dark-red violet	[22]
S	Sun color	[23]
Leaf morphology and heading habit		
gl	Glossy foliage	[24]
sm	Smooth leaves, with wr (originally S)	[25]
Pet	Petiolate leaves vs. sessile	[26]
W	Wide vs. narrower leaf	[26]
K	Dominant factor for heading	[27]
Plant habit T	Tall plant vs. short	[26]
Flower color and morphology		
Wh	White petal, dominant, wh, yellow	[28]
cp	Crinkly petal	[29]
cr	Cream petal, recessive	[29]
Cytoplasmic male sterility		
ms	Cytoplasmic factor with R, radish cytoplasm	[30]
Ms	Cytoplasmic factor with N, nigra cytoplasm	[28]
Self-incompatibility		
S	Self-incompatibility, multiple alleles	[31]
Disease resistance		
pb1, pb2	Duplicate genes, double recessive for resistance to *P. brassicae* race 6	[32]
F	Duplicate genes, double recessive for resistance to *P. brassicae* race 6	[33]
f	Major gene for resistance to black rot	[33]

Table 1.
Gene list of different characteristics of cabbage.

length, three chromosomes of intermediate length, and one chromosome of relatively short length [18]. The composition of the c genome of *B. oleracea* was ABBCCDEEF [19]. Cabbage, classified as a subspecies of *B. oleracea*, is a secondary polyploid with a fundamental chromosome number of 6 [20]. Three of the fundamental chromosomes are present in duplication, and the other nine are single [21].

The most valuable genes for geneticists and plant breeders are those that can be easily recognized and inherited in simple mendelian ratio because such genes can serve as markers (**Table 1**).

4. The breeding objectives pertaining to cabbage cultivation

- **High, consistent crop production:** Crop yield, the longstanding primary objective for breeders, remains a significant target, albeit with an increased emphasis on quality and additional parameters among most breeders.

- **Quality:** Crisp compact heads, a core length within half the head height, good flavor, and high nutrient content are all qualities a new variety should have.

- **Disease resistance:** Clubroot (*Plasmodiophora brassicae*), black rot (*Xanthomonas campestris* pv. *campestris*), fusarium yellows (*Fusarium oxysporum* f. *conglutinans*), downy mildew (*Peronospora parasitica*), turnip mosaic virus (TuMV), powdery mildew (*Erysiphe polygoni* D.C.), bottom rot (*Rhizoctonia solani* Kuhn) and tip burn are the most common diseases affecting cabbage, it is crucial to produce varieties that are resistant to many diseases.

- **Resistance to early bolting:** Cabbage is one of the most widely eaten vegetables at this time of year. Especially in years with a cold current in the spring, early bolting can result in significant economic losses for growers and breeders, highlighting the need for the breeding of early bolting resistant cultivars amenable to spring cultivation.

- **Cold and heat tolerance:** Cabbage must be bred for both cold and heat tolerance so that it may provide a high yield and mature quickly.

- **Storage ability:** The ability to store food for later consumption is a significant trait in cabbage, as is a compact head for ease of transport from the field to the market in the winter.

- **Consistency:** while sowing cabbage seeds at different times of the year, breeding early, mid, and late types completes the collection of cabbage variations, allowing for year-round harvesting.

- **Uniformity of the crop:** A consistent brassica field speeds up the grading process and cuts down on labor costs. The end goal is to have a field that is uniform in quality after being harvested just once.

- **Appearance:** The trait of color and shape holds significant importance in terms of appearance. The vegetable market has undergone a swift transformation due to advancements in packing and display facilities, resulting in a wide availability of commodities throughout the year. Additionally, the increased presence of visually appealing fruit and salad vegetables has compelled brassica producers to elevate their standards of presentation and quality.

5. Breeding methods

Due to the existence of sporophytic self-incompatibility, which results in heterozygous populations, cabbage is essentially a cross-pollinating plant that is mostly pollinated by bees and flies. Different population improvement schemes include mass selection, family breeding, recurrent selection, hybridization followed by selection in segregating generations, backcrossing, and heterosis breeding are breeding techniques that can be used to generate improved open pollinated types. In cabbage, selfing can be carried out for up to 3–4 generations without considerable inbreeding depression; however, the genotype affects the degree of inbreeding depression. Bud pollination is used for selfing one to two days before anthesis [34].

6. The systems for the development of hybrids

6.1 Male sterility

6.1.1 Genie male sterility

The occurrence of male sterility in *B. oleracea* is determined by a single gene, ms, which is a recessive trait resulting from a mutation of the male fertile gene, Ms. [9, 35]. Male sterile plants possess the ability to produce female fertile offspring, albeit with slightly smaller flowers and anthers compared to their male fertile counterparts.

6.1.2 Cytoplasmic male sterility

The presence of this characteristic was identified in Japanese radish [30]. CMS is caused by the interaction between nuclear and mitochondrial genomes. However, there is no fertility restorer gene present in any Japanese radish cultivar [36, 37]. Through numerous backcrosses, the cabbage nucleus was successfully inserted into the cytoplasm of the Ogura male sterile radish [38]. Regrettably, it was observed that all the genotypes of oleracea that were introduced into the cytoplasm of this radish displayed varying degrees of yellowing (chlorosis) in the young leaves when the seedlings were cultivated at temperatures of 12°C or lower [38]. The level of yellowing exhibited variation depending on the selection of recurring male parents. Two cytoplasmic male sterile cabbage germplasms through a breeding process involving the crossing of *Raphanus* (radish) and *B. oleracea*, followed by seven generations of repeated backcrossing with cabbage [39].

Nonetheless, both cabbage lines exhibited a slight manifestation of chlorosis, a condition that was resolved by means of protoplast fusion. In a study, a standardized protocol for the fusion of protoplasm in order to facilitate the transfer of desirable male sterility cytoplasm from broccoli to cabbage established [40]. Many male sterile lines created in a research using an improved cytoplasmic male sterile line known as R3625 that consistently displayed sterility while retaining normal growth and development [34]. presented A new cabbage hybrid, named Qiugan No.1, in their study [41]. This hybrid was developed by crossing the CMS line CMS02 with the inbred line 97,025-B. With consistent sterility, good adaptability, and great combining ability, 16 cabbage CMS lines [42]. The F1 hybrid Qiugan No. 4, a cabbage variety, was developed by [43] through the crossbreeding of the CMS line CMS95100 with the SI line 98,017-3-5-6-5-2. The compact head had a weight of 2 kilograms. The present study aimed to investigate the impact of Ogura cytoplasm introgression on various quality traits in a total of 17 cabbage lines belonging to the species *Brassica oleracea* var. *capitata* L. [44]. The transfer of Ogura cytoplasm was accomplished by introducing cytoplasmic male sterility (CMS) from the EC-173419 source line of cabbage obtained from the National Bureau of Plant Genetic Resources in New Delhi, India. This transfer was achieved by repeated backcrossing for a minimum of six generations, resulting in the incorporation of Ogura cytoplasm into various nuclear backgrounds of cabbage. The experimental findings demonstrated that the introduction of Ogura cytoplasm had a significant impact on various quality traits. Overall, there was an observed increase of 3–5 times in the concentration of various nutritional compounds in certain lines, whereas a reduction of 4–5 times was observed in others. The introgression of *Diplotaxis catholica* and *Trachystoma ballii* male sterile cytoplasms into cabbage background was carried out during the summer season of 2019. This was achieved through the utilization of backcrossing and embryo rescue technique [45].

6.2 Self-incompatibility

Given that cabbage sticky pollen is not dispersed by wind, it can be inferred that cabbage flowers rely on insect-mediated cross-pollination. The self-compatibility of plants varies depending on the cultivar, with some displaying self-incompatibility while others exhibit self-compatibility. Self-incompatibility serves as a mechanism to hinder self-fertilization, as well as the fertilization process in crosses involving plants with identical genotypes or nearly identical genotypes, wherein the expression of incompatibility specificity is identical due to dominance or co-dominance. The control of incompatibility specificity in *brassicas* is governed by a single locus known as the S gene [46, 47]. An elucidation of the self-incompatibility mechanism in *B. oleracea*, focusing on its cellular and molecular aspects provided [48].

According to the study conducted by [49], it was observed that *Brassica oleracea* typically exhibits a substantial number of S-locus alleles, often reaching up to 50. The initial cabbage hybridization occurred in Japan in 1950, employing self-incompatible lines. The hybrid was identified as Nagaoka No.1. Cabbage hybrids are currently generated through self-incompatibility mechanisms; however, a transition to male sterility is anticipated, likely to occur by the year 1997.

The degree of self-incompatibility can be evaluated by quantifying the quantity of seeds generated following either self-pollination or cross-pollination. The process of bud pollination as a mechanism to overcome self-incompatibility involves the deliberate opening of the bud and subsequent transfer of pollen from an open flower of the same plant [50]. The style and stigma of the flower become fully receptive approximately 3–4 days prior to its opening, during which time the self-incompatibility factor has not yet manifested. Consequently, self-fertilization becomes feasible, allowing for the circumvention of self-incompatibility [51, 52]. Currently, there exist two alternative approaches that have gained significant traction in addressing the issue of self-incompatibility [46]. The initial step entails the application of a sodium chloride solution with a concentration of 3–4% onto the exposed blossom. The flower is subsequently allowed to remain undisturbed for a duration of 20–30 minutes. Any surplus salt solution is eliminated by either blowing it off the flower or absorbing it with a moist cloth or paper towel. Following this, the process of self-pollination can take place. The application of salt facilitates the removal of the inhibitor present in the stigma, thereby enabling the process of self-pollination. An alternative and more effective approach, particularly when dealing with a large number of plants, entails the self-pollination of the exposed flowers, followed by the placement of the plants within an enclosed facility or chamber that allows for the controlled introduction of 5% carbon dioxide. Subsequently, the self-incompatibility mechanism is surmounted, thereby facilitating the generation of seeds subsequent to self-pollination [53–55].

7. Heterosis breeding

In order to successfully breed hybrid cultivars, three key prerequisites must be met. Firstly, a significant level of heterosis must be present. Secondly, a robust system for producing a large quantity of hybrid seeds must be in place. Lastly, an effective method for identifying hybrids with exceptional combining ability is essential [56, 57].

The phenomenon of heterosis in cabbage yield was initially observed by [28]. It was observed that there was a significant level of heterosis ranging from 30–100% in terms of yield in cabbage [58]. The hybrid combination of Enkhuizen and Valvatevka exhibited enhanced vigor, a more condensed growth habit, and increased productivity. According to [9, 59] findings, there was a notable presence of heterosis in relation to yield. In the diallel analysis conducted by [15], a total of six parents and 15 F1 hybrids were utilized. The findings of this study demonstrated the presence of heterosis, specifically favoring the superior parents in terms of the marketability of heads and net weight. A varietal cross between Copenhagen market and Pusa Drum Head demonstrated notable heterosis in relation to yield, net weight, and compactness [15].

The phenomenon of heterosis was investigated, revealing that hybrids exhibited increased plant weight accompanied by a relatively reduced number of outer leaves compared to their parental counterparts [60]. In evaluating the performance of 45 F1 hybrids for cabbage, [61] discovered considerable heterosis for head weight relative to superior parents in 22 hybrids. The heterosis for head bulk and homogeneity in single and double cross F1 hybrids in contrast to their double haploid inbred lines [62].

In order to create the F1 cabbage hybrid Zhonggan 192, [63] crossed the early maturing CMS- 87-534 (Cytoplasmic Male Sterile Line) with the inbred line 88-62-1-1. It takes 60 days from transplant to harvest. The cross combinations, namely SI I-4-6 × Glory-7, SI III-I-I × KGAT-1, IIS CMS × Glory-7, IIS CMS × KGAT-1, and SI III-I-I × E-1-1 & 2, exhibited notable positive heterosis compared to the control groups, indicating their high potential [64].

8. Resistance breeding

The development of resistant varieties is widely regarded as the most viable and environmentally sustainable approach for disease management. Resistance breeding has yielded numerous instances of success in diverse vegetable crops. However, the cultivation of resistant varieties necessitates a comprehensive comprehension of the evolutionary interplay between the host and the pathogen. The efficacy of resistance breeding hinges upon a comprehensive comprehension of the genetic origins of resistance, the racial composition of the pathogen, and the genetic underpinnings of the host-pathogen interaction. Understanding the scope of manipulation within the host-pathogen interaction is also crucial (**Tables 2–4**). The sources of R genes in plant breeding can include advanced breeding lines, newly developed genetic stocks through pre-breeding efforts, commercial varieties, landraces or primitive cultivars, and wild relatives in the form of original progenitors or related species. Additionally, one method of incorporating a resistance gene into a breeding program involves the incorporation of a resistant parent through hybridization. Moreover, the backcross technique is widely employed in breeding to introduce resistance traits into pre-existing adapted cultivars. This approach does not disrupt the overall genetic makeup of the recipient commercial variety. The transfer of monogenic dominant resistance to downy mildew and black rot into cultivated varieties can be achieved through the utilization of the backcross method. Moreover, the incorporation of a single R parent with favorable horticultural characteristics in the process of hybrid breeding can lead to the development of resistant hybrids against these specific pathogens. The gene pyramiding approach can be utilized to develop varieties that possess resistance to both the Kalia and Singh diseases [65, 66].

Diseases	Sources
Black Rot	MR-I, Invento (F_1), 83-6, Pusa Mukta (Sel.8), AC 204, AC 208, RRM, RC-20
Alternaria	MR-I, AC 204, 83-6, 83-2, 83-1, RRM, 561, 563
Downy Mildew	MR-I, AC-204, 83-6, 83-2, AC-208, CC-10
Aphid (*Brevicoryne brassicae*)	All season, Red Drum Head, Sure Head, Express Mail

Table 2.
Sources identified In cabbage in India.

MR-I	Black Rot, Downy Mildew, *Sclerotinia* Rot, *Alternaria*, Diamond Back Moth
AC 204	Black Rot, *Alternaria*, Downy Mildew
RRM	Diamond Back Moth, *Sclerotinia* Rot, Black Rot

Table 3.
Multiple resistance sources identified in cabbage.

Disease	Pathogen
Bacterial	
Black Rot	*Xanthomonas campestris* pv. *campestris* (Pammel) Dowson
Spot of Cabbage	*Pseudomonas cichorii* (Swingle) Staff
Club Root	*Plasmodiophora brassicae* War.
Fungal	
Downy Mildew	*Perenospora parasitica*
Leaf Spot and Blight	*Alternaria brassicae* and *A. brassiciola*
Yellows or Fusarium Wilt	*Fusarium oxysporum* f. sp. *conglutinans*
Black Leg	*Phoma lingum*
Insect and Pest	
Diamondback Moth	*Plutella xylostella* L.
Cabbage Stem Borer	*Hellula undalis* Fab.
Cabbage Caterpillar	*Pieris brassicae* L.
Cabbage Semi-loopers	*Plusia orichalcea* Fab. and *P. nigrisigna* Walker
Aphids	*Brevicoryne brassicae* L.

Table 4.
Important disease and insect Pest of cabbage.

9. Breeding for heat tolerance

Heat stress resistance refers to the differential performance exhibited by certain genotypes when exposed to equivalent levels of heat stress. The mechanisms associated with heat stress can be classified into two main categories: heat avoidance and heat tolerance. The ability of a genotype to dissipate radiation energy and prevent an increase in plant architecture to a stress level is indicated by heat avoidance. Six quantitative trait loci (QTLs) on chromosomes 2, 4, and 6 that are associated with resistance to head splitting. Two quantitative trait loci (QTLs), namely SPL-2-1 and SPL-4-1, are situated on chromosomes 2 and 4, correspondingly [67].

The markers BRPGM0676 and BRMS137 were found to exhibit a strong linkage with the trait of head-splitting resistance. Furthermore, these markers were observed to be conserved within the QTL SPL-2-1 region. During the months of October and November, an evaluation was conducted on two cabbage genotypes, PA-1 and PA-2, which are classified as 'No-chill type'. The evaluation focused on determining the genotypes' performance in terms of earliness and head yield. Among the tested varieties, PA-2 exhibited promising results in terms of head yield, achieving a yield of 22 metric tons per hectare. Additionally, PA-2 displayed desirable head traits suitable for November maturity [45].

10. Transgenics in cabbage

Genetic transformation includes two methods to transfer foreign genes into plants: *Agrobacterium*mediated gene transfer or the indirect method (vector mediated) and direct gene transfer (vectorless) method (**Table 5**). Through *Agrobacterium tumefaciens*-mediated transformation with Bacillus thuringiensis (Bt) cry genes, Jin et al. [70] targeted resistance to diamondback moth larvae in cabbage.

Cultivar	Technique of gene transfer	Gene transfer	Improvement in traits	References
King Cole	*Agrobacterium tumefaciens*	cry1a (c)	Resistance to diamond back moth	[68]
Yingchun, Jingfeng	*A. tumefaciens*	CpTI	Insect tolerance to *Pieris rapae* L.	[69]
Scorpio, Testie	*A. tumefaciens*	cry1Ia3	Resistance to diamond back moth	[70]
Uji	*A. tumefaciens*	GO	Enhanced tolerance to black rot	[71]
DTC 507	*A. tumefaciens*	cry1 b	Resistance to diamond back moth	[72]
Summer Summit, KY cross	Biolistic method	cry1Ab	Resistance to diamond back moth	[73]
A21-3	*A. tumefaciens*	cry1Ia8, cry1Ba3	Resistance to diamond back moth	[74]
Pride of India	*A. tumefaciens*	cry1Aa	Resistance to diamond back moth	[75]
Golden Acre	*A. tumefaciens*	betA	Enhanced salt tolerance	[76]
KY Cross	*A. tumefaciens*	AtHSP101	Increase the high temperature tolerance	[77]
AD BENTAM	*A. tumefaciens*	JMT	Resistance to heat stress	[78]

Table 5.
Transgenics for resistance.

All cabbage plants that were genetically modified with a synthetic Bt gene, specifically cry1Ab3, exhibited complete resistance to larvae of the pest. *Agrobacterium*-mediated transformation to introduce a synthetic fusion gene derived from *Bacillus thuringiensis* into a tropical cabbage breeding line known as 'DTC 507' [72]. This fusion gene encoded a translational fusion product consisting of Cry1B and Cry1Ab δ-endotoxins. In their study, Russell et al. [79] conducted a comprehensive review on the advancements made in the utilization of pyramided Bt genes cry1B and cry1C for managing populations of *Plutella xylostella*, *Crocidolomia pavonana*, *Hellula undalis*, and *Pieris* spp. in cabbage.

11. Marker-assisted breeding

Marker-assisted selection (MAS) is a method in which the selection process is conducted based on a marker rather than the actual trait itself. The effective implementation of Marker-Assisted Selection (MAS) hinges upon the strong correlation between the marker and the primary gene or Quantitative Trait Locus (QTL) accountable for the specific trait [80]. The utilization of novel genomic tools expedites the process of identifying markers that are closely associated with specific genomic regions. The phenomenon of combining genetic material from various sources to confer resistance against a common disease is an illustrative instance and represents one of the most prevalent implementations of gene pyramiding. The utilization of molecular markers increases the probability of detecting the presence of a genotype that combines advantageous alleles within a population [81]. The utilization of contemporary molecular techniques has emerged as a significant factor in comprehending the arrangement and connections of the *Brassica* genomes. The findings from these studies not only validated the source of the amphidiploid species, but also indicated that the A and C genomes can be traced back to a common lineage, while the B genome is genetically distinct from both the A and C genomes, forming an independent lineage [82]. The genome of Cole, a member of the *Brassica oleracea* species [83]. This study also identified several molecular markers that are associated with significant traits in various *Brassica oleracea* crops. Genomic sequences serve as valuable resources in the creation of reliable DNA markers. The genetic maps in *Brassica* have a dual purpose: (a) to comprehend the interrelation between the genomes of the cultivated diploid species of *Brassica*, and (b) to facilitate the application of genetics and breeding techniques in the cultivation of various *Brassica* crops (**Tables 6** and 7).

Trait	Gene	Marker	References
Genic male sterility	CDMs399-3	EST-SSR	[84]
Head shape	QTLs (Htd 3.1, Htd 8.1)	SSR, InDel	[67]
Head splitting	QTLs (SPL-2-1, SPL-4-1)	SSR	[67]
Yellow-green leaf	Ygl-1	InDel	[83]
Plant height, leaf length, head transverse diameter	QTLs Ph 3.1, Ll 3.2, Htd 3.2	InDel	[85]

Table 6.
Linked molecular markers in cabbage.

Source	Trait	Gene	Marker	Reference
DH line 'Reiho P01'	Black rot-resistant	LG2 and LG9	CAPS and SRAP	[89]
Line 'Early Fuji'	Black rot-resistant	C2, C4 and C5	SNP	[90]
C1234 inbred line	Black rot resistant	BRQTL-C3, BRQTL-C6	SNP-based CAPS	[91]
Glossy leaf cabbage (748)	DBM resistance	—	SSR	[92]
Brassica oleracea var. capitata 96–100 inbred line	Head-splitting resistance	SPL-2-1, SPL-4-1	SSR and InDel	[93]
B. oleracea var. capitata BN1 variety	High temperature and high humidity tolerant	—	SNPs	[94]

Table 7.
Molecular marker studies conducted for resistance to biotic and abiotic stresses.

12. CRISPR/Cas9 in cabbage

The utilization of Clustered Regularly Interspaced Short Palindromic Repeats (CRISPR)/CRISPR-associated protein 9 (Cas9) has emerged as a contemporary technique for manipulating genomes [86]. The system comprises a nuclease known as Cas9, along with two short single-stranded RNAs, namely crRNA and tracrRNA. These two RNAs are combined to create a single-guide RNA (sgRNA), which is utilized for the purpose of genome editing. The Cas9 protein, in conjunction with a guide RNA (gRNA), combines to create a ribonucleo protein complex that subsequently attaches to the genomic DNA. The application of CRISPR-Cas9 technology for gene editing in *Brassica* vegetables. The study employed cabbage as the model plant and targeted the PsbS gene through the creation of a Brassica deletion mutant [86]. The utilization of preassembled ribonucleoprotein complexes (RNPs) in the introduction of these complexes into cabbage protoplasts using PEG 4000 as a facilitator [87]. Four single guide RNAs (sgRNAs) were utilized to target two specific endogenous genes, namely the FRI and PDS genes. Each gene was targeted by two sgRNAs. The introduced sgRNAs were subsequently assessed, and a mutation rate ranging from 1.15 to 24.51% was observed. In their study, [88] employed CRISPR/Cas9 gene editing to specifically target three genes in cabbage, namely phytoene desaturase gene (BoPDS), S-receptor kinase gene (BoSRK), and male-sterility-associated gene BoMS1. To facilitate this gene editing process, the researchers utilized a construct containing tandemly arrayed tRNA-sgRNA architecture. According to their findings, the mutation in the BoSRK3 gene resulted in the complete suppression of self-incompatibility, while the mutation in the BoMS1 gene led to the development of a fully male-sterile mutant.

13. Conclusions

Through the utilization of traditional breeding methods, substantial advancements have been made in the cultivation of numerous cabbage cultivars, which exhibit adaptability to diverse climatic and cultural environments. The task of producing cabbage with increased yield, improved quality, and enhanced nutritional value poses a significant challenge for plant breeders from the standpoint of consumer demand. The cultivars and breeding lines that exhibit resistance to black rot are distinct and should be perpetuated as valuable sources of novel germplasm for plant breeders. While the presence of glucosinolates in cruciferous crops remains a subject of concern, efforts will be made to develop cabbages not only for their culinary properties or adaptability to culture and management, but also for their therapeutic values. Contemporary methodologies, in conjunction with traditional breeding techniques, employ wild species and their relatives to effectively overcome diverse biotic and abiotic stresses in order to facilitate the creation of novel varieties. In forthcoming times, it will be imperative to prioritize the enhancement of nutritional characteristics, fortification against biotic and abiotic pressures, and the development of herbicide resistance using contemporary biotechnological methodologies. To date, limited endeavors have been directed towards enhancing cabbage through the utilization of advanced technologies such as transgenic methods and genome editing techniques like CRISPR/Cas9. These technologies are crucial for harnessing the potential of introducing genes that confer resistance to pests and diseases. Despite limited progress, it is imperative to address concerns regarding biosafety and environmental safety pertaining to transgenic crops. Furthermore, it is crucial to fully exploit the capabilities of contemporary biotechnological methods for enhancing the genetic traits of cabbage.

Author details

Shipra Singh Parmar[1]*, Impa H. Ravindra[2] and Ramesh Kumar[1]

1 Department of Vegetable Science, Dr. Y.S. Parmar University of Horticulture and Forestry, Solan, HP, India

2 Department of Vegetable Science, College of Horticulture, University of Horticultural Sciences, Bagalkot, Karnataka, India

*Address all correspondence to: shipraparmar690@gmail.com

References

[1] Ordás A, Cartea ME. Cabbage and kale. In: Vegetables. New York: Springer; 2008. pp. 119-149. DOI: 10.1007/978-0-387-30443-4_4

[2] Munger H. Adaptation and breeding of vegetable crops for improved human nutrition. In: Quebedeaux B, Bliss FA, editors. Horticulture and Human Health. Englewood Cliffs, NJ: Prentice Hall; 1988. pp. 177-184

[3] Heyland DK, Johnson AP, Reynolds SC, Muscedere J. Procalcitonin for reduced antibiotic exposure in the critical care setting: A systematic review and an economic evaluation. Critical Care Medicine. 2011;**39**(7):1792-1799. DOI: 10.1097/CCM.0b013e31821201a5

[4] Singh BK, Sharma SR, Singh B. Heterosis for superoxide dismutase, peroxidise and catalase enzymes in the head of single cross hybrids of cabbage (*Brassica oleracea* L. var. *capitata* L.). Journal of Genetics. 2010;**89**(2):217-221. DOI: 10.1007/s12041-010-0028-8

[5] Williams DJ, Edwards D, Hamernig I, Jian L, James AP, Johnson SK, et al. Vegetables containing phytochemicals with potential antiobesity properties: A review. Food Research International. 2013;**52**:323-333

[6] Thamburaj S, Singh N. Text Book of Vegetables, Tuber Crops and Spices. New Delhi: ICAR Publication; 2001 99p

[7] Food and Agriculture Organization. FAOSTAT. 2019. Available from: http://www.fao.org/foastat/en/#data/QC

[8] National Horticulture Board. First Advance Estimate of Area and Production of Cabbage in India; 2019. Available from: http://nhb.gov.in

[9] Nieuwhof M. Cole crops. London: World crop ser. Leonard Hill; 1969. p. 353

[10] Phillips R, Rix M. The Random House Book of Vegetables. NY: Random House; 1993

[11] Mitchell ND. The status of *Brassica oleracea* L. subsp. *oleracea* (wild cabbage) in the British Isles. Watsonia. 1976;**11**:97-03

[12] Boswell VR. Our vegetable travellers. National Geographic. 1949;**96**:145-117

[13] Ordás A, Cartea ME. Cabbage and kale. In: Prohens J, Nuez F, editors. Vegetables I. Handbook of Plant Breeding. Vol. 1. New York: Springer; 2008. pp. 119-149. DOI: 10.1007/978-0-387-30443-4_4

[14] Hodgkin T. Cabbages, kales etc. *Brassica oleracea* (Cruciferae). In: Smartt J, Simmonds NW, editors. Evolution of Crop Plants. London: Longman; 1995. pp. 76-82

[15] Swarup V, Brahmi P. Cole crops. In: Dhillon BS, Tyagi RK, Saxena S, Randhawa GJ, editors. Plant Genetic Resources: Horticultural Crops. New Delhi: Narosa Publishing House; 2005. pp. 75-88

[16] Chiang BY, Grant WF, Chiang MS. The somatic karyotype of cabbage (*Brassica oleracea* ssp. *capitata*). Euphytica. 1979;**28**(1):41-45. DOI: 10.1007/BF00029171

[17] UN. Genomic analysis in Brassica with special reference to the experimental formation of *B. napus* and peculiar mode of fertilization. Journal of Japanese Botany. 1935;7:389-352

[18] Sikka SM. Cytogenetics of Brassica hybrids and species. Journal of Genetics. 1940;**40**(3):441-474. DOI: 10.1007/BF03028231

[19] Roebbelen G. Contributions to the analysis of the Brassica-genome. Chromosoma. 1960;**11**:205-228. DOI: 10.1007/BF00328652

[20] Kamala T. Basic chromosome number and the probable origin of the genomes in *Brassica*. Current Science. 1978;**47**:128-129

[21] Prakash S. Haploidy in *Brassica nigra* Koch. Euphytica. 1973;**22**(3):613-614. DOI: 10.1007/BF00036663

[22] KristofFerson JB. Contribution to the genetics of *Brassica oleracea*. Hereditas. 1924;**5**:297

[23] Magruder R, Myers CH. The inheritance of some plant color in cabbage. Journal of Agricultural Research (Washington, DC). 1933;**47**:233

[24] Moore JF, Anstey TH. A study of the degree of natural selfing in green sprouting broccoli (*B. oleracea* L. var. *italica*). Proceedings of the American Society for Horticultural Science. 1954;**63**:440-442

[25] Kwan C. Inheritance of some plant characters in cabbage, *Brassica oleracea* var. *capitata*. Journal of The Agricultural Association of China. 1934;**126-127**:81

[26] Pease MS. Genetic studies in *Brassica oleracea*. Journal of Genetics. 1926;**16**(3):363-385. DOI: 10.1007/BF02983007

[27] Allgayer H. Genetic investigations with garden cabbage by crossing trials by Richard Freendenberg. Z Indukt Abstamm Vererbungsl. 1928;**47**:191

[28] Pearson OH. The influence of inbreeding upon the season of maturity in cabbage. Proceedings of the American Society for Horticultural Science. 1932;**29**:359

[29] Wills AB. A preliminary gene list in *Brassica oleracea*. EUCARPIA Cruciferae Newsletter. 1977;**2**:222

[30] Ogura H. Studies on the new male sterility in Japanese radish with special reference to the utilization of this sterility towards the practical raising of hybrid seed. Memoirs of the Faculty of Agriculture, Kagoshima University. 1968;**6**:9

[31] Odland ML, Noll CJ. The utilization of cross-compatibility and self-incompatibility in the production of F_1 hybrid cabbage. Proceedings of the American Society for Horticultural Science. 1950;**55**:390

[32] Chiang MS, Crête R. Inheritance of clubroot resistance in cabbage (*Brassica oleracea* L. var. *capitata* L.). Canadian Journal of Genetics and Cytology. 1970;**12**(2):253-256. DOI: 10.1139/g70-036

[33] Williams PH, Staub J, Sutton JC. Inheritance of resistance in cabbage to black rot. Phytopathology. 1972;**62**(2):247. DOI: 10.1094/Phyto-62-247

[34] Fang Z, Liu Y, Lou P, Liu G. Current trends in cabbage breeding. Journal of New Seeds. 2005;**6**:75-107

[35] Rundfeldt H. Untersuchungen zur zuchtung des Kopfkohls (*B. oleracea* L. var. *capitata*). Z. Pflanzenzucht. 1960;**44**:30

[36] Jian YC, Ding YH. Breeding and utilization of *Brassica oleracea*

cytoplasmic male sterile lines. China Vegetables. 2005;**6**:4-6

[37] Shiga T. Male sterility and cytoplasmic differentiation. In: Tsunoda S, Hinata K, Gomez-Campo D, editors. Brassica Crops and Wild Allies, Biology and Breeding. Vol. 205. Tokyo: Japan Science Society Press; 1980

[38] Bannerot HO, Boulidard L, Chupeau Y. Unexpected difficulties met with the radish cytoplasm in Brassica oleracea. Eucarpia Cruciferae Newsletter. 1977;**2**:16

[39] McCollum GD. CMS (ESG) 508 and CMS (ESG) 512 cytoplasmic male-sterile cabbage germplasm (with radish cytoplasm). Horticultural Science. 1988;**23**(1):227-228. DOI: 10.21273/HORTSCI.23.1.227

[40] Sigareva MA, Earle ED. Direct transfer of a cold-tolerant Ogura male-sterile cytoplasm into cabbage (*Brassica oleracea* ssp. *capitata*) via protoplast fusion. Theoretical and Applied Genetics. 1997;**94**(2):213-220. DOI: 10.1007/s001220050402

[41] Jian YC, Ding YH, Qu GQ. A new cabbage F1 hybrid-'Chungan No. 2'. China Vegetables. 2008;**2**:35-37

[42] Ding Y, Jian Y. Cabbage cytoplasmic male sterile line breeding and its application in hybrid seed production. Acta Horticulturae. 2008;**771**:89-95

[43] Kang JG, Ding Y, Jian YC. A new autumn cabbage F_1 hybrid 'Qiugan No. 4'. China Vegetables. 2010;**20**:74-76

[44] Parkash C, Kumar S, Singh R, Kumar A, Thakur N, Kumar S, et al. Introgression of 'Ogura'cytoplasm in cabbage alters its nutritional quality and antioxidant activities. Zemdirbyste-Agriculture. 2019;**106**(3): 273-280

[45] IARI [Annual Report]. Vol. 110. New Delhi: Indian Agricultural Research Institute, India; 2019. p. 012

[46] Dixon GR. Vegetable Brassicas and Related Crucifers. Wallingford, UK: CA B International; 2007

[47] Wallace DH. Procedures for identifying S-allele-genotype of *Brassica*. Theoretical and Applied Genetics. 1979;**54**(6):249-265. DOI: 10.1007/BF00281207

[48] Nasrallah ME, Wallace DH. The influence of modifier genes on the intensity and stability of self-compatibility in cabbage. Euphytica. 1968;**17**:493

[49] Brace J, Ryder CD, Ockendon DJ. Identification of S-alleles in *Brassica oleracea*. Euphytica. 1994;**80**(3):229-234. DOI: 10.1007/BF00039654

[50] Dickson MH, Wallace DH. Cabbage breeding. In: Bassett MJ, editor. Breeding Vegetable Crops. Vol. 395. Westport Connecticut: AVI Publishing Co. Inc.; 1986

[51] Prakash C, Verma TS, Kumar PR. Genetic analysis of cabbage using self-incompatible lines. Indian Journal of Agricultural Sciences. 2003;**73**(3):412-413

[52] Rai N, Singh AK. Genetic variability, heritability and genetic advance studies in cabbage. Journal of Applied Biology. 2000;**10**(1):8-11

[53] Haruta T. Studies on the Genetics of Self- and Cross-Incompatibility in Cruciferous Vegetables. Minneapolis: Northrop King and Co.; 1966 67 p

[54] Kumar PR, Yadav SK, Sharma SR, Lal SK, Jha DN. Impact of climate change on seed production of cabbage in North Western Himalayas. World Journal of Agricultural Sciences. 2009;**5**(1):18-26

[55] Parkash C, Dey SS, Bhatia R, Dhiman MR. Indigenously developed SI and CMS lines in hybrid breeding of cabbage. Indian Journal of Horticulture. 2015;**72**(2):212-217. DOI: 10.5958/0974-0112.2015.00041.9

[56] Kibar B, Karaagac O, Kar H. Heterosis for yield contributing head traits in cabbage (*Brassica oleracea* L. var. *capitata* L.). Ciencia e Investigación Agraria. 2015;**42**:205-216

[57] Pathak S, Kumar J, Vidyasagar VS. Heterosis and combining ability for marketable yield and component traits in cabbage (*Brassica oleracae* L. var. *capitata* L.). Indian Journal of Agricultural Sciences. 2007;**77**:97-100

[58] Khimic PE. Heterosis in cabbage. Plant Breeding. 1939;**Abstr 9**:518

[59] Tanaka N, Niikura S. Genetic analysis of the developmental characteristics related to the earliness of head formation in cabbage (*Brassica oleracea* L. var. *capitata* L.). Breeding Science. 2006;**56**(2):147-153. DOI: 10.1270/jsbbs.56.147

[60] Nieuwhof M, Garretsen F. Hybrid breeding in early spring cabbage II. Euphytica. 1975;**24**:87-97

[61] Moore TA, Wallace DH. Combining ability and heterosis using self-incompatible lines in cabbage (*Brassica oleracea* var. *capitata* L.). The Indian Journal of Genetics and Plant Breeding. 1988;**47**:20

[62] Kaminski P. The evaluation of head cabbage [*Brassica oleracea* L. var. *capitata* L.] F1 hybrids derived by the use of DH lines. Vegetable Crops Research Bulletin. 2001;**54**(1):13-18

[63] Zhuang M, Fang ZY, Liu YM, Yang LM, Zhang YY, Sun PT. A new spring cabbage hybrid 'Zhonggan 192'. Acta Horticulturae Sinica. 2010;**37**:1881-1882

[64] Kumar N, Chadha S, Kanwar S. CMS and SI based heterosis for yield and related traits in low chill cabbage under mid hills condition of Himachal Pradesh. Indian Journal of Horticulture. 2019;**76**(4):663-671. DOI: 10.5958/0974-0112.2019.00106.3

[65] Janssen K, Schöfl G, Reineke A, Heckel DG, Groot AT. Oviposition of diamondback moth in the presence and absence of a novel host plant. Bulletin of Entomological Research. 2011;**101**(1):99-105

[66] Kalia P, Singh S. Accelerated improvement of cole vegetable crops. In: Gosal S, Wani S, editors. Accelerated Plant Breeding. Vol. 2. Springer Nature Switzerland; 2020. pp. 101-135

[67] Pang W, Li X, Choi SR, Nguyen VD, Dhandapani V, Kim Y, et al. Mapping QTLs of resistance to head splitting in cabbage (*Brassica oleracea* L.var. *capitata* L.). Molecular Breeding. 2015;**35**(5):126. DOI: 10.1007/s11032-015-0318-1

[68] Metz TD, Dixit R, Earle ED. *Agrobacterium tumefaciens* mediated transformation of broccoli (*Brassica oleracea* var. *italica*) and cabbage (*B. oleracea* var. *capitata*). Plant Cell Reports. 1995;**15**(3-4):287-292. DOI: 10.1007/BF00193738

[69] Hongjun F, Dali L, Guanlin W, Yinghui L, Zhen Z, Xianghui L. An insect-resistant transgenic cabbage plant with cowpea trypsin inhibitor

(CpTI) gene. Acta Botanica Sinica. 1997;**39**(10):940-945

[70] Jin R-G, Liu Y-B, Tabashnik BE, Borthakur D. Development of transgenic cabbage (*Brasica oleracea* var. *capitata*) for insect resistance by *Agrobacterium tumefaciens*- mediated transformation. In Vitro Cellular & Developmental Biology. 2000;**36**(4):231-237. DOI: 10.1007/s11627-000-0043-1

[71] Lee YH, Yoon IS, Suh SC, Kim HI. Enhanced disease resistance in transgenic cabbage and tobacco expressing a glucose oxidase gene from *aspergillus Niger*. Plant Cell Reports. 2002;**20**(9):857-863. DOI: 10.1007/s00299-001-0416-x

[72] Paul A, Sharma SR, Sresty TVS, Devi S, Bala S, Kumar PS, et al. Transgenic cabbage (*Brassica oleracea* var. *capitata*) resistant to diamondback moth (*Plutella xylostella*). Indian Journal of Biotechnology. 2005;**4**:72-77

[73] Liu CW, Lin CC, Yiu JC, Chen JJW, Tseng MJ. Expression of a bacillus thuringiensis toxin (cry1Ab) gene in cabbage (*Brassica oleracea* L. var. *capitata* L.) chloroplasts confers high insecticidal efficacy against *Plutella xylostella*. Theoretical and Applied Genetics. 2008;**117**(1):75-88. DOI: 10.1007/s00122-008-0754-y

[74] Yi D, Cui L, Wang L, Liu Y, Zhuang M, Zhang Y, et al. Pyramiding of Bt cry1Ia8 and cry1Ba3 genes into cabbage (*Brassica oleracea* L. var. *capitata*) confers effective control against diamondback moth. Plant Cell, Tissue and Organ Culture (PCTOC). 2013;**115**(3):419-428. DOI: 10.1007/s11240-013-0373-4

[75] Gambhir G, Kumar P, Aggarwal G, Srivastava DK, Thakur AK. Expression of cry1Aa gene in cabbage imparts resistance against diamondback moth (plutella xylostella). Biologia Futura. 2020;**71**(1-2):165-173. DOI: 10.1007/s42977-020-00014-8

[76] Bhattacharya RC, Maheswari M, Dineshkumar V, Kirti PB, Bhat SR, Chopra VL. Transformation of *Brassica oleracea* var. *capitata* with bacterial betA gene enhances tolerance to salt stress. Scientia Horticulturae. 2004;**100**(1-4):215-227. DOI: 10.1016/j.scienta.2003.08.009

[77] Rafat A, Aziz MA, Rashid AA, Abdullah SNA, Kamaladini H, Sirchi MHT, et al. Optimization of *Agrobacterium tumefaciens*-mediated transformation and shoot regeneration after cocultivation of cabbage (*Brassica oleracea* subsp. *capitata*). Scientia Horticulturae. 2010;**124**(1):1-8. DOI: 10.1016/j.scienta.2009.11.015

[78] Hur SH, Min BW. Efcient development of transgenic cabbage with jasmonic acid carboxyl methyltransferase (JMT) gene based on PMI/mannose selection system. Plant Breeding. 2015;**3**(3):226-237

[79] Russell DA, Huang D, Bhalla P, Singh M, Robin C, Golz J, et al. Progress in the development of transgenic cabbage, cauliflower and canola expressing stacked bts for caterpillar control and RNAi for aphid suppression. The Mysore Journal of Agricultural Sciences (Special Issue) A. 2017;**51**:159-167

[80] Osei MK, Ruth JA. Marker-assisted selection (MAS): A fasttrack tool in tomato breeding. In: Recent Advances in Tomato Breeding and Production. London, UK, London: IntechOpen; 2018. pp. 93-113

[81] Ishii T, Yonezawa K. Optimization of the marker-based procedures for pyramiding genes from multiple

donor lines: II. Strategies for selecting the objective homozygous plant. Crop Science. 2007;**47**(5):1878-1886. DOI: 10.2135/cropsci2006.11.0750

[82] Warwick SI, Black LD. Molecular systematics of Brassica and allied genera (subtribe Brassicinae, Brassiceae) chloroplast genome and cytodeme congruence. Theoretical and Applied Genetics. 1991;**82**(1):81-92. DOI: 10.1007/BF00231281

[83] Liu XP, Yang C, Han FQ, Fang ZY, Yang LM, Zhuang M. Genetics and fine mapping of a yellow-green leaf gene (ygl-1) in cabbage (*Brassica oleracea* var. *capitata* L.). Molecular Breeding. 2016;**36**:1-8

[84] Chen C, Zhuang M, Fang Z, Wang Q, Zhang Y, Liu Y, et al. A codominant marker BoE332 applied to marker-assisted selection of homozygous male-sterile plants in cabbage (*Brassica oleracea* var. *capitata* L.). Journal of Integrative Agriculture. 2013;**12**(4):596-602. DOI: 10.1016/S2095-3119(13)60277-4

[85] Lv H, Wang Q, Liu X, Han F, Fang Z, Yang L, et al. Whole-genome mapping reveals novel QTL clusters associated with main agronomic traits of cabbage (*Brassica oleracea* var. *capitata* L.). Frontiers in Plant Science. 2016;7:989. DOI: 10.3389/fpls.2016.00989

[86] Jansson S. Gene-edited plants on the plate: The 'CRISPR cabbage story'. Physiologia Plantarum. 2018;**164**(4): 396-405. DOI: 10.1111/ppl.12754

[87] Murovec J, Guček K, Bohanec B, Avbelj M, Jerala R. DNA-free genome editing of *Brassica oleracea* and *B. rapa* protoplasts using CRISPR-Cas9 ribonucleoprotein complexes. Frontiers in Plant Science. 2018;**9**:1594. DOI: 10.3389/fpls.2018.01594

[88] Ma C, Zhu C, Zheng M, Liu M, Zhang D, Liu B, et al. CRISPR/ Cas9-mediated multiple gene editing in *Brassica oleracea* var. *capitata* using the endogenous tRNA-processing system. Horticulture Research. 2019;**6**(1):20. DOI: 10.1038/s41438-018-0107-1

[89] Doullah MAU, Mohsin GM, Ishikawa K, Hori H, Okazaki K. Construction of a linkage map and QTL analysis for black rot resistance in *Brassica oleracea* L. Indian Journal of Natural Sciences. 2011;**1**:1-6

[90] Kifuji Y, Hanzawa H, Terasawa Y, Ashutosh N, Nishio T. QTL analysis of black rot resistance in cabbage using newly developed EST-SNP markers. Euphytica. 2013;**190**(2):289-295. DOI: 10.1007/s10681-012-0847-1

[91] Lee J, Nur KI, Murukarthick J. Genome-wide SNP identification and QTL mapping for black rot resistance in cabbage. BMC Plant Biology. 2015;**15**:1-11

[92] Ramchiary N, Pang W, Nguyen VD, Li X, Choi SR, Kumar A, et al. Quantitative trait loci mapping of partial resistance to diamondback moth in cabbage (*Brassica oleracea* L). Theoretical and Applied Genetics. 2015;**128**(6):1209-1218. DOI: 10.1007/s00122-015-2501-5

[93] Su Y, Liu Y, Li Z. QTL analysis of head splitting resistance in cabbage (*Brassica oleracea* L. var. *capitata*) using SSR and indel makers based on whole-genome re-sequencing. PLoS One. 2015:9

[94] Song H, Lee M, Hwang B, Han C, Park J, Hur Y. Development and application of a PCR-based molecular marker for the identification of high temperature tolerant cabbage (*Brassica oleracea* var. *capitata*) genotypes. Agronomy. 2020;**10**(1):116. DOI: 10.3390/agronomy10010116

Chapter 6

Breeding for Macronutrient Use Efficiency (NTUE) in Legumes

Vaishnavi Vijayakumar, Sumaiya Sulthana, Balaji Kannan, Sivakumar Rathinavelu, Anandhi Krishnan and Rajaprakasam Sudhagar

Abstract

Increasing population warrants increasing food crop productivity with a minimum input cost. The usage of inorganic fertilizers is inevitable in modern agriculture cropping systems. Nitrogen, phosphorus, and potassium are the major nutrients used by agriculturalists worldwide since the effect of these nutrients is highly significant on crop productivity; therefore, it is crucial to use them in an optimized way to make farming economically sustainable. The capacity of crops to absorb nutrients and efficiently utilize them is known as nutrient use efficiency (NtUE). The NtUE of plants would reduce the usage of synthetic fertilizers, reduce the nutrient leaches into the environment, and increase crop productivity. The development of optimum biomass-producing and nutrient-efficient crop varieties are the key for rational agriculture.

Keywords: legumes, food security, macronutrients, nutrient use efficiency, plant breeding strategies

1. Introduction

To cater to the calorie requirement of the exploding population, especially in Asia, where legumes continue to be the main source of protein, the thrust to develop better-yielding, nutrient-rich, resource-use efficient, and stress-tolerant cultivars is increasing. Fertilizer application is required to augment the nutrient status of soils because the majority of arable soils are low in important nutrient(s). Most legumes are grown as rainfed crops in Asia, where inputs including water are used either in a minimal amount or not at all. Over time, after successful domestication and subsequent crop improvement activities, the nutrient use efficiency of legumes is altered. Enhanced NtUE in legume varieties is considered crucial, which otherwise reduces the production costs. Cultivars with high NtUE will protect the environment by minimizing the need for synthetic/inorganic fertilizers, lowering the rate of nutrient leaching into the ecosystems, increasing yield, and ensuring agricultural sustainability while preserving the quality of the soil and groundwater. The plant has the capacity to absorb and store nutrients in significant quantities. The presence of sufficient quantities of nutrients helps the plants combat the challenges posed by various stresses. Further, genotypes with good NtUE expand the area under legume production, which is otherwise classified as minimal-nutrient-available marginal lands. These lands are nevertheless amenable to cultivation. Genetic improvement for NtUE therefore is considered an important target in legume breeding because it has the potential to significantly enhance production and sustain the effects of climate change.

The perspective of NtUE varies according to the discipline. At each level, the input and the intended output vary, and different terminologies are used. Agro-ecologists, agronomists, and physiologists/breeders/biotechnologists are the major discipline of scientists involved in the NtUE research. Agroecologists perceive the agricultural system, encompassing associated ecosystems, as having a specific NtUE in relation to the agricultural goods it produces. Agronomists focus on the field and concentrate on factors such as yield, which is influenced by applied seeds; fertilizers; pesticides; and other inputs. They also include manpower costs for the specific area of land. Plant breeders/physiologists/biotechnologists, in their perspective, consider that the plant has an inherent NtUE, which is influenced by varied internal/external stimuli. In this chapter, we approach the NtUE from a breeder/physiologist/biotechnologist perspective.

2. Roles of macronutrients on plant performance

The nutrients nitrogen (N), potassium (K), and phosphorus (P) regulate plant growth and metabolism in significant ways and therefore are called macronutrients. Nitrogen, a key constituent of nucleic acids, and alkaloids regulate protein and chlorophyll synthesis. N, in association with a few other minor nutrients, improves the water and nutrient absorption efficiency of roots [1]. Potassium (K) plays an important role in water regulation, osmotic balance, enzymatic activation, stomatal opening and closing, plant defense, stress tolerance, yield, and quality improvement. Phosphorus (P) regulates many plant functions: (i) energy transfer; (ii) photosynthesis; (iii) nutrient, sugar, and starch transport; and (iv) character inheritance (**Figure 1**).

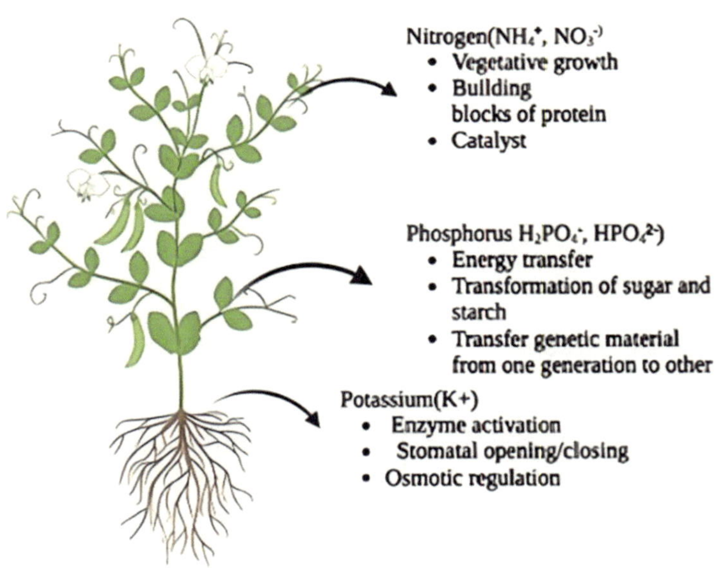

Figure 1.
Role of major nutrients in plants. (created with https://www.biorender.com).

3. Mechanisms to improve NtUE

The NtUE represents the ability of plants to absorb and utilize nutrients to maximize yield. The NtUE involves three crucial plant processes: nutrient absorption, assimilation, and utilization. The key factors determining the NtUE are presented in **Figure 2**.

For the effective utilization of applied and accessible nutrients, plants primarily employ a few strategies like (i) increased root growth, (ii) changing the rhizosphere's characteristics to increase nutrient availability, (iii) altering root architecture in all possible ways to harbor significant interactions between rhizosphere microbial populations, and (iv) regulating the expression of ion transporters.

4. Internal and external factors that affect nutrient use efficiency

The NtUE depends on both internal (plant) and external factors. Fertilizer-dependent factors, climate variables, microbial population, and other agronomic practices are examples of external factors. Internal factors include genetic, physiological, and biochemical regulations (**Table 1**).

5. Nitrogen use efficiency (NUE)

NUE is the productivity ability of plants in a unit of land with the available nitrogen (both naturally occurring and supplemented by N fertilizer) [2]. NUE is the result of (i) utilization efficiency (yield/absorbed N) and (ii) absorption efficiency

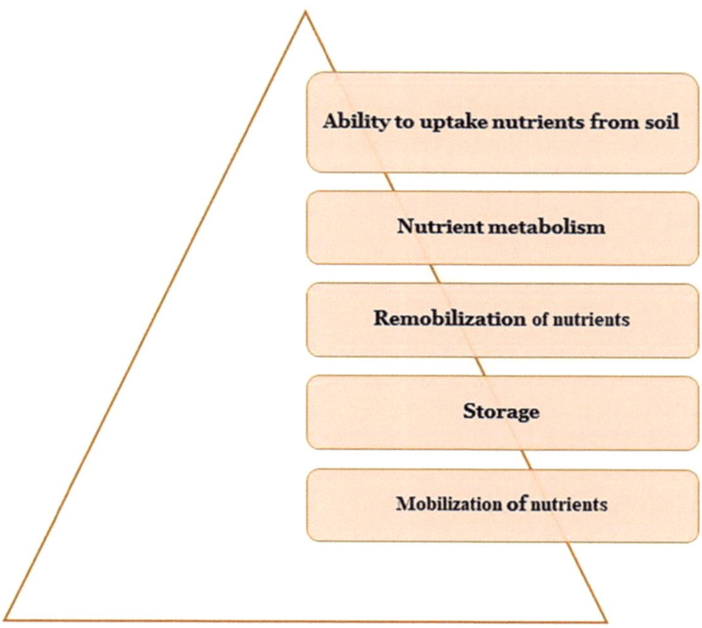

Figure 2
Factors determining NtUE in legumes.

Internal factors	External factors
1. Genetic factors: • Types of species/cultivar/genotypes 2. Physiological factors: • Length of the root • The density of root hair • Mobilization of nutrients within the plant • Presence of ion transporters. 3. Biochemical factors: • Availability of enzymes like nitrate reductase (N), phosphatase (P), *etc*. for the conversion of nutrients • Root exudates: citric acid, malic acid, and trans-aconitic acid	1. Fertilizer-dependent factors: • Source of fertilizer • Time of application • Depth and method of application • When applied in combination • Use of slow-release form. 2. Climatic factors: • Soil moisture • Temperature • Rainfall • Erosion 3. Microbial population: • Arbuscular mycorrhizae and other beneficial soil microbes 4. Others: • Incorporated crop residues • Crop rotation especially for legumes • Cover crops • Weed control measures, diseases, and insects.

Table 1.
Factors influencing NtUE.

(it is the ratio between absorbed and available N). Knowledge of the different forms of nitrogen that are preferred by the plants for uptake, enzymes involved in the conversion processes, transporters involved in N transport, and gene expression studies will help the breeders to formulate effective strategies to improve the NUE.

Forms of N and their conversion: N fertilizer supplies different forms of nitrogen upon soil application, and such forms are narrated in **Table 2**. Regardless of the source of N applied to soil, plants prefer the nitrate form. Excess application of N forms leads to certain environmental concerns (**Figure 3**).

Enzymes involved in N assimilation: Plants with the help of certain transporters absorb these different forms of nitrogen, and a few enzymes support plants to use them. Nitrate reductase, nitrogenase, nitrite reductase, glutamate synthase, and glutamine synthetase are the primary enzymes associated with the reduction of nitrate and assimilation of ammonia. Ammonium is the main source of inorganic N in plants, which is converted

	Organic N	Ammonium N	Nitrate N
Plant Use	Not used	Usable	Usable
Loss through	Mineralization, erosion	Volatilization, erosion	Leaching and denitrification
Changes	Mineralization	Immobilization, nitrification	Immobilization, denitrification

Table 2.
Different forms of nitrogen.

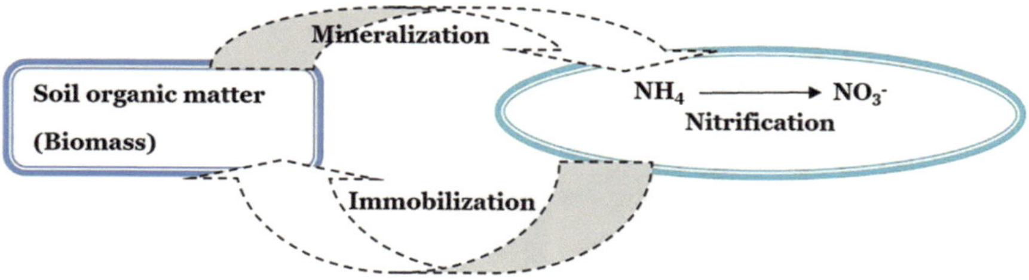

Figure 3.
Conversion of different forms of nitrogen.

into the organic molecules glutamine and glutamate by the enzymes GS (Glutamine synthase) and GOGAT (Glutamine synthetase-glutamate synthase). In plants, GS and GOGAT are present in several isoenzyme forms and distinctively expressed depending on the developmental stage of cells. The GS enzyme has two different forms: (i) a cytosolic form (GS1) found in phloem cells, roots, and leaves and (ii) a plastidic form (GS2) present in the chloroplast and mitochondria tissues used for photosynthesis. Each GS isoform plays unique roles in photo-respiratory, NH3 (ammonia) assimilation, $No3^-$ (nitrate) reduction, N translocation, and recycling. Variations in the proportions of GS1 and GS2 are noticed within plant organs and between plants. The enzyme GOGAT has two isoforms, each of which plays a specific role in the primary assimilation or recycling of N. A ferredoxin-dependent isoenzyme (Fd-GOGAT) is primarily involved in the re-assimilation of photo-respiratory ammonia, working in tandem with GS2. Glutamate synthesis (in photosynthetic and/or non-photosynthetic tissues) is regulated by a pyridine nucleotide-dependent isoform (NADH-GOGAT) (**Figure 4**).

N transporters: To adapt to changes in soil N, plants must have highly developed absorption and signaling mechanisms. After the absorption and conversion of nutrients to available forms by the plants, the next important process will be the proper distribution of available nutrients to the plant parts for their conversion into biomass and grains (from source to sink) [3]. Certain transporters will also regulate this distribution of nutrients within the plants. Nitrate transporters are primarily responsible for nitrate absorption and translocation to the needy tissues [2, 4]. Transport and portioning of N are aided by xylem and phloem cells, respectively. Nitrate activates (i) transporter family genes like NRT1 and NRT2 and (ii) the assimilation pathway genes nitrate reductase (NR) and nitrite reductase (NiR) (**Figure 5**) [5].

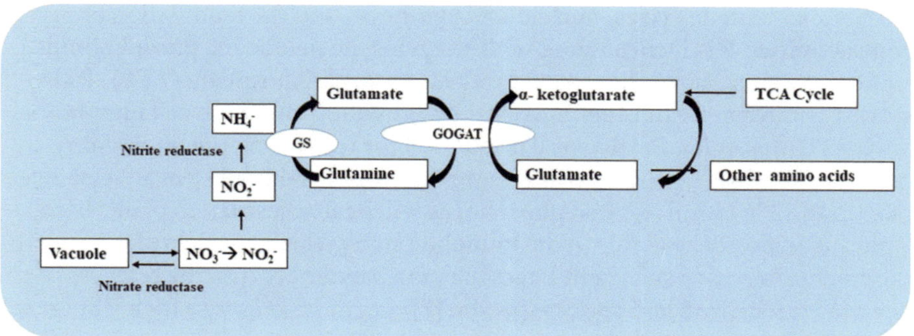

Figure 4.
Nitrate assimilation.

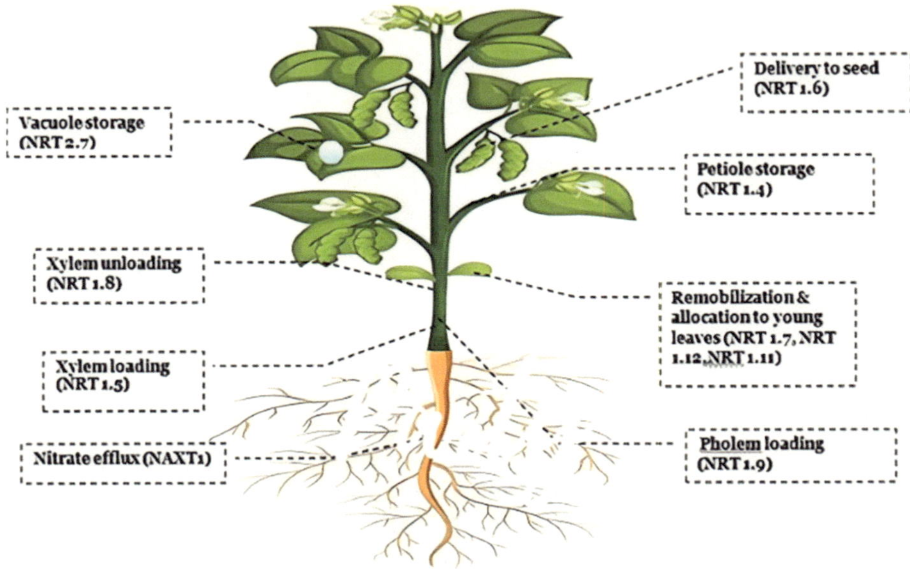

Figure 5.
Types of N transporters. (image adapted from https://www.freepik.com/).

Gene expression studies: Nitrate activates a few genes very quickly (within minutes) without the need for protein synthesis, which is referred to as "Primary Nitrate Response" (PNR) [6]. A number of kinases, transcription factors, and non-coding RNAs [4] control PNR. Nitrate is a signaling molecule that is essential for the development and growth of plants. Legumes have an advantage over non-legume plants in that they can fix nitrogen through the symbiotic interaction between N-fixing diazotrophs, requiring less nitrogen fertilizer [5]. Following the identification of a structural and/or regulatory gene connected to metabolic pathway regulation, information can be gained by creating overexpressers or choosing mutants with the gene of interest. Because legume leaves contain more nitrogen (N), there is higher photosynthesis as a result.

6. Phosphorus use efficiency (PUE)

Phosphorus (P) is the macronutrient that regulates plant growth by its association with several metabolic cycles, nucleic acid synthesis, and the modulation of several enzyme activities. P is incorporated with enzymes, nucleic acids, phospholipids, phosphor-proteins, sugar phosphates, and adenosine triphosphate (ATP). P also supports the movement of other nutrients by providing energy to cell membranes through ATP molecules. Sufficient P is required for root development and its enhanced functions (**Figure 6**), which support crop sustainability in adverse situations. Up to 80% of soil-applied phosphorus will be inaccessible to plants because phosphorus is mostly available in the immobile state, which restricts plant growth and development. Access to sufficient P fertilizer is a barrier because the reserves of rock phosphate are localized and region-specific [7]. Legumes require a high P because symbiotic N fixation in nodules consumes energy, and root nodules also function as strong P sinks. Low P causes a 30–40% growth and yield reduction [8]. The absorption of P by a plant is facilitated by root hairs, tips, the outermost layers of root cells,

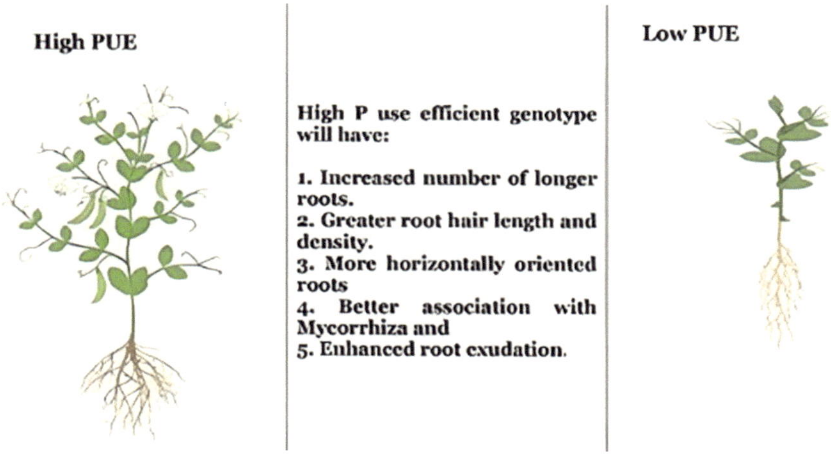

Figure 6.
Comparison of plant growth under high and low P availability conditions. (created with https://www.biorender.com).

and the root-colonized mycorrhizal fungi. The absorbed P is either stored in the root or transported and stored in other cells based on the requirement.

In low soil P situations, to achieve a higher PUE, legumes modify their root morphology, organic acid exudation, associations with mycorrhizal fungi, and signaling mechanisms [8]. P-use efficiency mainly depends on two processes P acquisition efficiency and P utilization efficiency. P acquisition efficiency mainly depends on the ability of the roots to absorb P from the soil so it can be regulated by changing the root morphological characteristics, architecture, and organic acid exudation. The major modifications plants accomplish for a better PUE are associated with root modifications and transporter deployment, which are discussed hereunder.

Root morphological modifications: Root morphological alterations related to length, volume, density, shallowness, exudates, and hair influence the PUE. For example, to achieve efficient P absorption under the P deficit situations, the primary root growth is stimulated [9]. To ensure a greater Pi acquisition, the lateral roots proliferate and expand soil exploration [10], and such modifications are based on soil Pi availability [11]. Changes in the root epidermal cell-borne hairs aid in greater absorption of nutrients, and their role in significant Pi uptake is well documented in alfalfa [5, 12]. Cluster root, which is one of the modifications of root otherwise called proteinoid roots, also helps in P acquisition. Cluster root has numerous dense short rootlets, which increase PUE through (i) increasing the surface area between soil and plant and (ii) secreting a good quantity of phosphatases, organic acids, and protons. In legumes, this type of root modification is mainly observed in white lupin [13]. Root architecture is another factor that determines PUE [14]. The legume root developmental pattern depends on soil P availability. Under P-deficiency situations, legumes develop shallow roots to cope with the mineral stress [15]. Shallow roots with a greater lateral rooting are observed in soybean under P deficiency, and such modifications increase Pi uptake more than deep root systems [16], and these observations are also well documented in a cultivated soybean core collection [16]. Soil P is highly immobile; therefore, it needs to be converted to inorganic form for proper absorption by plants. This conversion can be accelerated by the root exudates like organic acids, phosphatases, and a few others. Various organic acids are secreted from

Transporters	PHT1	PHT2	PHT3	PHT4
Location	Plasma membranes	Chloroplast membranes (plastids)	Mitochondrial membranes	Golgi apparatus
Function	Helps in P acquisition under low and high P situations	Helps in P mobility	Helps in an effective P distribution	Helps in P movement between cytosols

Table 3.
Details of P transporter subfamilies.

different types of legumes like pigeon pea, soybean, common bean, and so on. [17]. For example, white lupin produces cluster roots that secrete malate and citrate, which help in increasing Pi concentration.

P transporters: Four subfamilies of P transporters, PHT1, PHT2, PHT3, and PHT4, regulate Pi uptake and transport in plants. On the basis of structural differences and subcellular localization, these regulators are classified into different subfamilies [18] (**Table 3**).

Gene expression studies: Organic acid synthesis in legumes is regulated by a few genes. Among them, phosphodiesterase or hydrolase enzymes, particularly purple acid phosphatases (PAPs), play key roles in hydrolyzing organic P. The resultant product of hydrolyzation is inorganic Pi, which is readily absorbed by plants. The expression studies of these genes can be targeted for increased P uptake.

7. Potassium use efficiency (KUE)

Potassium (K+) makes up about 10% of a plant's dry weight [19] and is considered a vital macronutrient. K+ has numerous roles in cell physiology and metabolism. It increases root growth, improves drought resistance, activates enzymes, maintains turgor in cells, enhances translocation of sugars and starch, and imparts resistance against several biotic stresses [20]. The plants growing in K^+ limited environments are stunted and underdeveloped. K^+ also influences the maturity of various crops in different ways.

Potassium absorption and mobility: Plants maintain a required level of potassium content within. To achieve this, plants use various mechanisms for sensing the availability of K+ in the soil, but these mechanisms are yet to be deciphered [21]. In K+-sufficient conditions, the external K+ is transported into the plant *via* plasma membrane through several channels and transporters. The K+ uptake and utilization comprise three processes: (i) root uptake, (ii) in & out movement in the xylem/ phloem, and (iii) cellular compartmentalization [22]. In K+-deficit conditions, plants expand their root systems to ensure sufficient absorption through enhanced root surface contact with soil. An increase in root volume is achieved by over-expressing transcription factors that are responsible for root formation. Plants use two systems of K+ uptake depending on the K+ availability [23]. When the available soil K+ is greater than 100 mM, plants utilize a low-affinity K+ uptake system where the inward rectifying channels are shaker-type K+. High affinity K+ uptake is used when the available soil K+ is less than 100 mM [11]. Plant Cyclic Nucleotide Gated Channels (CNGC) are the type of non-selective channels that are also employed by plants in certain situations; however, the reasons for such utilization are not well known [24]. Generally, in plants, for achieving an effective K+ transport, K+ uptake permeases (KT/HAK/KUP), the Trk/HKT transporter family, and the CPA (cation proton antiporters) are used [25]. After absorption, K+ is transported to the needy tissues through loading/unloading channels of the xylem and phloem. In Arabidopsis roots, the first shaker-like outward rectifying K+ channel 'Stelar K+ Outward Rectifier' (SKOR) is reported, which is associated with K+ transport from stellar cells to the xylem [26]. Intracellular movement of K+ is achieved through sequestration into vacuoles and guard cell movement. These movements are regulated by several K+ transporters and channels [27].

8. Breeding for nutrient use efficiency (NtUE)

The focus has been on enhancing NtUE in legumes due to scarce nutrient sources and environmental issues in real field conditions. Nutrient deficiency stress can cause an adaptive reaction, such as increased root volume for improved access to nutrients. For increasing the nutrient use efficiency of crops, certain conditions need to be fulfilled. They include the presence of useful genetic variation, information on gene action and trait heritability, and an appropriate knowledge of physiological determinants of NtUE. Detailed information is to be generated on nutrition acquisition and utilization, for formulating breeding strategies to achieve an efficient NtUE. The nutrient acquisition involves (i) changes in the rhizosphere properties (both biological and chemical), (ii) enhanced root growth, (iii) changes in the root architecture, (iv) microbial interactions in the rhizosphere, and (v) expression and regulation of ion transporters. An effective utilization can be achieved through greater root-to-shoot nutrient translocation, compartmentalization, partitioning, and remobilization [28]. From the research findings available on public platforms, the following breeding strategies are suggested to enhance the NtUE (**Figure 7**).

To ensure success in the NtUE programs, an integrated approach involving physiological and biochemical methods, classical breeding approaches, novel breeding methods, and omics strategies are to be formulated. Germplasm exploration, characterization, cataloging, and quantifying the existing genetic variability for NtUE are the basic plant breeding steps to be accomplished, which will further help in formulating the type of advanced research to be considered. When the required quantum of variability does not exist in the germplasm, induced mutagenesis can be employed to

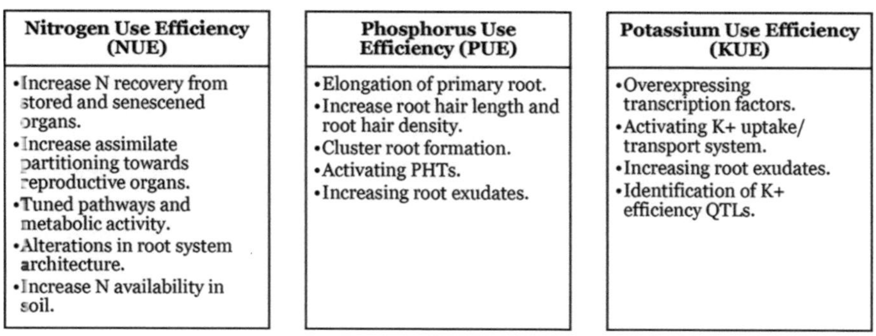

Figure 7.
Strategies to be targeted to enhance the NtUE.

evolve variability. Thereon, MutMap can be considered for tagging genes or QTLs. In the situation of sufficient variability for NtUE, the genotypes can be screened under sub-optimal conditions of the nutrients. After the identification of specific genetic stocks, either MAGIC population development or classical hybridization followed by pedigree selection can be considered. When nutrient-specific donors are identified, mapping populations can be developed for the identification of QTLs and their further utilization. Different genome editing strategies can be employed to normalize various regulators (positive and negative) associated with NtUE. The CRISPR/dead (d) Cas9 (dCas9)-aided promoter engineering can be employed for cytosine/adenine base editing. To achieve the over-expression of targeted genes, this system uses transcriptional factors (repressors/activators) [29]. Different 'omics' strategies can be employed to understand the genetic architecture of NtUE and their consequent effects.

9. Conclusion

An adequate supply of critical nutrients in balanced ways is the key to increasing crop yield. As farmers strive to increase production and maintain profitability in the face of rising fertilizer costs and changing climate conditions, NtUE is considered important. In a production system where the food crops are nutrient-starved, improving soil fertility using fertilizers is necessarily required to sustain productivity. To achieve this requirement, synthetic fertilizers were used in enormous quantities after the Green Revolution. Indiscriminate use of synthetic fertilizers leads to environmental pollution and, therefore, warrants better management of the soil nutrient equilibrium where NtUE offers scope to achieve this target. However, progress has been slow due to the complexity of soil and plant genomes affecting crop nutrition, a lack of complete knowledge of the genetics of NtUE, and an inconsistent conceptual understanding of NtUE. With a better understanding of NtUE, scientists can make significant progress in creating more nutrient-efficient cultivars with the use of different cutting-edge technologies.

The mechanisms to increase NtUE of the plant species may vary according to nutrients and crops. The following concluding remarks shall be considered to achieve an effective NtUE.

NUE: One of the significant N sources in farming is obtained through biological nitrogen (N_2) fixation. Therefore, it is suggested to identify genotypes with enhanced biological nitrogen-fixing ability that can otherwise help in effective NUE. The varieties developed in the past 50 years were bred and/or selected under high mineral fertilization inputs, which otherwise narrowed the chance of using genetic variations under low mineral fertilization conditions, thus warranting the inception of targeted selection programs. The best-performing genotypes for varied N situations shall not be selected based on the interactions between genotypes and levels of N since lines that perform best at high/ideal N levels may not be promising well at deficit/low N level(s). Evolution of new plant varieties that (i) absorb more N from the soil irrespective of form and (ii) effective utilization of absorbed N. Breeding for more active rhizobial and Arbuscular mycorrhizal (AM) symbioses in legumes can be an intriguing alternative to boosting plant yield.

PUE: Phosphorus, is highly immobile in the soil; therefore, proper P absorption is very difficult. Increasing PUE is therefore achieved by increasing root volume, root architecture, root exudates, and transporters. Significant variations are identified

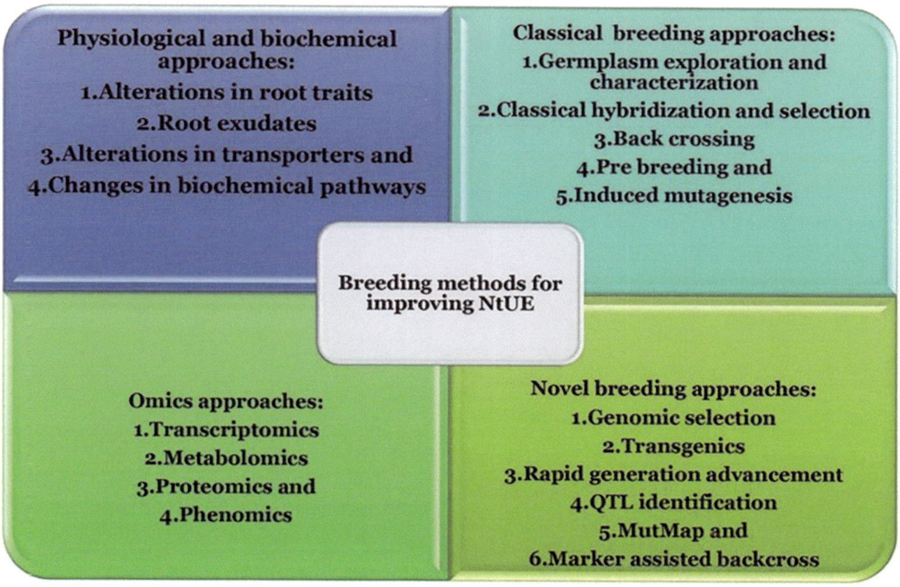

Figure 8.
Different breeding methodologies for improving NtUE in legumes.

for different mechanisms to improve PUE in legumes. Development of transgenic soybean plants with over-expressed Gm6PGDH1 ensures greater Pi uptake under P-deficient conditions through significant lateral root growth [16]. Based on a RIL-based QTL analysis in beans under P-deficit situations, it is concluded that the QTL for basal root growth angle is co-segregated with grain yield and whose further introgression would facilitate more PUE [19]. Genome-wide association studies can be helpful in identifying important genes for improving PUE. Other methods, such as alternate gene splicing, overexpression of certain genes, and utilizing long noncoding RNA, have also been used to enhance PUE nutrient response in crops.

KUE: Development of mapping population and QTL identification, genome broadening, hereditary and evolutionary analysis, and gene mapping and identification are the commonly used techniques to improve KUE breeding. The genes for low-K tolerance and good shoot dry weight are identified *via* QTL analyses and GWAS, which are considerably facilitated by large amounts of genome-wide re-sequencing data, and high-quality reference genomes can further be exploited. Transcriptomic information offers insight into gene expression patterns in response to various K treatments; as a result, it has been widely used to research the genes, biological processes, and metabolic processes that determine low K tolerance. Using transcriptome profiling and QTL mapping together, it is possible to find potential genes that regulate low K tolerance. Proteomics, metabolomics, and ionomics investigations on K consumption efficiency in crops have been conducted, but they are significantly less prevalent than transcriptome research. Alternative mRNA splicing (AS) and noncoding RNAs are the main topics of epigenomics studies. The nutshell of the above-described methodologies is presented in **Figure 8** for better understanding of the readers.

Author details

Vaishnavi Vijayakumar[1], Sumaiya Sulthana[1], Balaji Kannan[2], Sivakumar Rathinavelu[3], Anandhi Krishnan[4] and Rajaprakasam Sudhagar[5*]

1 Centre for Plant breeding and Genetics, Tamil Nadu Agricultural University (TNAU), Coimbatore, Tamil Nadu, India

2 Department of Physical Sciences and Information Technology, TNAU, Coimbatore, Tamil Nadu, India

3 Department of Crop Physiology, CMS, TNAU, Coimbatore, Tamil Nadu, India

4 Department of Pulses, CPBG, TNAU, Coimbatore, Tamil Nadu, India

5 CPBG, TNAU, Coimbatore, Tamil Nadu, India

*Address all correspondence to: sudhagar.r@tnau.ac.in

References

[1] Andrews M, Lea PJ, Raven JA, Azevedo RAD. Nitrogen use efficiency. 3. Nitrogen fixation: Genes and costs. The Annals of Applied Biology. 2009;**155**(1):1-13

[2] Tegeder M, Masclaux-Daubresse C. Source and sink mechanisms of nitrogen transport and use. The New Phytologist. 2018;**217**(1):35-53. DOI: 10.1111/nph.14876

[3] Zhang B, Meng S, Gong M. The expected and unexpected roles of nitrate transporters in plant abiotic stress resistance and their regulation. International Journal of Molecular Sciences. 2018;**19**(11):9-11. DOI: 10.3390/ijms19113535

[4] Krouk G, Crawford NM, Coruzzi GM, Tsay YF. Nitrate signaling: Adaptation to fluctuating environments. Current Opinion in Plant Biology. 2010;**13**(3):265-272

[5] Wang YY, Cheng YH, Chen KE, Tsay YF. Nitrate transport, signaling, and use efficiency. Annual Review of Plant Biology. 2018;**69**:85-122

[6] Hao QN, Zhou XA, Sha AH, Wang C, Zhou R, Chen SL. Identification of genes associated with nitrogen-use efficiency by genome-wide transcriptional analysis of two soybean genotypes. BMC Genomics. 2011;**12**:1-15

[7] Smit AL, Bindraban PS, Schröder JJ, Conijn JG, Van der Meer HG. Phosphorus in agriculture: Global resoources, trends and developments: Report to the steering committee technology assessment of the Ministery of agriculture, nature and food quality. In: The Netherlands, and in Collaboration with the Nutrient Flow Task Group (NFTG), Supported by DPRN (Development Policy Review Network). Wageningen: Plant Research International; 2009

[8] Chen Z, Wang L, Cardoso JA, Zhu S, Liu G, Rao IM, et al. Improving phosphorus acquisition efficiency through modification of root growth responses to phosphate starvation in legumes. Frontiers in Plant Science. 2023;**14**:1094157. DOI: 10.3389/fpls.2023.1094157

[9] Zhang H, Yang Y, Sun C, Liu X, Lv L, Hu Z, et al. Up-regulating GmETO1 improves phosphorus uptake and use efficiency by promoting root growth in soybean. Plant, Cell & Environment. 2020;**43**(9):2080-2094. DOI: 10.1111/pce.13816

[10] Williamson LC, Ribrioux SP, Fitter AH, Leyser HM. Phosphate availability regulates root system architecture in Arabidopsis. Plant Physiology. 2001;**126**(2):875-882. DOI: 10.1104/pp.126.2.875

[11] Zhou J, Xie J, Liao H, Wang X. Overexpression of b-expansin gene GmEXPB2 improves phosphorus efficiency in soybean. Physiologia Plantarum. 2014;**150**(2):194-204. DOI: 10.1111/ppl.12077

[12] Li C, Li K, Liu X, et al. Transcription factor GmWRKY46 enhanced phosphate starvation tolerance and root development in transgenic plants. Frontiers in Plant Science. 2021;**12**:700651. DOI: 10.3389/fpls.2021.700651

[13] Lynch JP, Brown KM. Topsoil foraging - An architectural adaptation of plants to low phosphorus availability. Plant and Soil. 2001;**237**(2):225-237. DOI: 10.1023/A:1013324727040

[14] Jha UC, Nayyar H, Parida SK, et al. Breeding and genomics approaches for improving phosphorus-use efficiency in grain legumes. Environmental and Experimental Botany. 2023;**205**:105120

[15] Pang J, Ryan MH, Lambers H, Siddique KH. Phosphorus acquisition and utilisation in crop legumes under global change. Current Opinion in Plant Biology. 2018;**45**:248-254

[16] Liang Q, Cheng X, Mei M, Yan X, Liao H. QTL analysis of root traits as related to phosphorus efficiency in soybean. Annals of Botany. 2010;**106**(1):223-234

[17] Ramaekers L, Remans R, Rao IM, Blair MW, Vanderleyden J. Strategies for improving phosphorus acquisition efficiency of crop plants. Field Crops Research. 2010;**117**(2-3):169-176

[18] Adams E, Shin R. Transport, signaling, and homeostasis of potassium and sodium in plants. Journal of Integrative Plant Biology. 2014;**56**(3):231-249. DOI: 10.1111/jipb.12159

[19] Shin R. Strategies for improving potassium use efficiency in plants. Molecules and Cells. 2014;**37**(8):575

[20] Gierth M, Mäser P. Potassium transporters in plants–involvement in K+ acquisition, redistribution and homeostasis. FEBS Letters. 2007;**581**(12):2348-2356

[21] Liu K, Li L, Luan S. Intracellular K+ sensing of SKOR, a shaker-type K+ channel from Arabidopsis. The Plant Journal. 2006;**46**(2):260-268

[22] Wang W, Zou J, White PJ, et al. Identification of QTLs associated with potassium use efficiency and underlying candidate genes by whole-genome resequencing of two parental lines in Brassica napus. Genomics. 2021;**113**(2):755-768

[23] He B, Hu F, Du H, et al. Omics driven crop potassium use efficiency breeding. Frontiers in Plant Science. 2022;**13**:1076193. DOI: 10.3389/fpls.2022.1076193

[24] Chen Y, Zhang S, Du S, Jiang J, Wang G. Transcriptome and metabonomic analysis of Tamarixramosissima potassium (K+) channels and transporters in response to NaCl stress. Genes. 2022;**13**(8):1313

[25] Fageria NK, Barbosa Filho MP, Da Costa JGC. Potassium-use efficiency in common bean genotypes. Journal of Plant Nutrition. 2001;**24**(12):1937-1945

[26] Li X, Zeng R, Liao H. Improving crop nutrient efficiency through root architecture modifications. Journal of Integrative Plant Biology. 2016;**58**(3):193-202. DOI: 10.1111/jipb.12434

[27] Le Thanh T, Hufnagel B, Soriano A, Divol F, Brottier L, Casset C, et al. Dynamic development of white lupin rootlets along a cluster root. Frontiers in Plant Science. 2021;**12**:738172

[28] McDonald G, Bovill W, Huang C, Lightfoot D. Nutrient use efficiency. In: Genomics and Breeding for Climate-Resilient Crops. Vol. 2. Target Traits. Germany: Springer; 2013. pp. 333-393

[29] Gajardo HA, GÃ3mez-Espinoza O, Boscariol Ferreira P, Carrer H, Bravo LA. The potential of CRISPR/Cas technology to enhance crop performance on adverse soil conditions. Plants. 2023;**12**(9):1892

Chapter 7

The Role of Metabolites in Abiotic and Biotic Stress Tolerance in Legumes

Soheila Afkar

Abstract

Population growth in the world has made the production of food to feed this population a major challenge. The Food and Agriculture Organization (FAO) estimates that to meet human food needs by 2050, crop productivity must double. Legumes family plays an important role in food security, poverty alleviation, and sustainability. It is determined that plant development and stress responses, as well as processes such as growth, the integrity of cells, energy storing, cellular signaling, formation of membrane and scaffolding, cellular replenishing, and whole-plant resource assignment, are managed by plant metabolites. One of the important parts of early stress responses concerns changes in plant metabolism, which includes the accumulation of antioxidants for the protection of cellular components from oxidative damage and the accumulation of compatible solutes that retain water in the cell. Other components, such as GABA and amino acids, including threonine, leucine, methionine, lysine, valine, and isoleucine, were usually induced during environmental stress conditions. In general, it was determined that plants containing various metabolites alter their physiology to adapt to various situations, such as stress. Important metabolites that play a role in tolerance to stress in legumes can help breeding programs in developing stress-tolerant cultivars to increase food security in the world.

Keywords: legumes, secondary metabolite, primary metabolite, protein, biotic stress, abiotic stress

1. Introduction

Concerning the increasing population of the world, the production of food to feed this population is a major challenge. The Food and Agriculture Organization (FAO) estimates that to meet human food needs by 2050, crop productivity must increase. However, because of climate change, the increase in production is not promising. In many countries, it is expected that the productivity of many crops will decrease by roughly 50% by 2080 [1, 2]. Changing global climatic conditions has a negative impact on food security and human health in two ways: 1- It directly affects the amount of food and indirectly affects the spread of disease and pests, water availability, and environmental pollution and 2- It changes CO_2 concentration that causes a change in plant biomass [1]. The third largest flowering family is the legume family, which is classified into five subfamilies, including Detarioideae, Cercidoideae, Dialioideae, Papilionoideae, Caesalpinioideae, and Duparquetioideae [2]. Due to their symbiotic relationship with nitrogen-fixing bacteria, the legume family is able to fix nitrogen, so they provide nitrogen for subsequent crops and reduce future nitrogen fertilizer usage [3, 4]. On the other hand, legumes are a family with a rich source of plant proteins, fibers, carbohydrates, dietary fibers, vitamins, and minerals, and they are low in fat [3–6]. Hence, seeds of the legume family have an important role in the human diet as they are rich in proteins, minerals, vitamins, and bioactive compounds [7]. Moreover, the legumes family plays an important role in food security, poverty alleviation, and sustainability [8]. Legumes provide 27% of global primary crop production and 33% of protein requirements and take the second place in food production after cereals [9]. In addition to primary metabolites, legumes are rich in secondary metabolites such as isoflavones, triterpenoid saponins, and inositol phosphates [5, 10, 11]. It is documented that stress tolerance in the legume family can be induced through the production of secondary metabolites such as phenolic, flavonoids, various alkaloids, and carotenoids [12].

Therefore, investigating the metabolites affecting the tolerance of biotic and abiotic stress in legumes helps researchers in breeding legumes to produce more yield under different stresses. In this part, with the aim of identifying important metabolites that play a role in tolerance to stresses in legumes, we reviewed metabolomics studies that focused on important abiotic stressors of legumes.

2. Drought stress

It is determined that water stress modifies the transcriptomic and metabolic profiles in plants. During drought, processes, such as the accumulation of compatible solutes and the production of hormones and antioxidants, occur for the regulation of the physiological processes. Also, the plant's ability to tolerate water stress is associated with soluble compounds such as polyamines, sucrose, trehalose, amino acids, polyhydric alcohols, mannose, and oligosaccharides. Therefore, the regulation of these compounds can be used for the improvement of drought tolerance in legume plants [13–15]. Several studies reported that changes in some key primary and secondary metabolites, such as sugars, amino acids, polyamines, phytohormones, secondary metabolites, antioxidants, and TCA cycle metabolites, are the response and adaptation mechanism of legumes to water-deficient conditions [16]. In a study by Goufo and colleagues, when cowpea (*Vigna unguiculata* L. Walp.) was exposed to water deficiency, 41 primary metabolites such as 5 sugars, 4 polyols, 24 amino acids, and 8 organic acids were detected. In addition, secondary metabolites such as Quercetin 3-O-6″-malonylglycoside, kaempferol 3-O-diglycoside, quercetin, and galactinol were identified [14]. Drought stress increased soybean plants' leucine, isoleucine, glycine, and proline. It seems that leucine, isoleucine, proline, and glycine are indicative of the potential for an advanced tolerance reaction under drought stress [17]. Cowpea plants responded to drought with an increase in quercetin 3-O-6′′-malonylglycoside and quercetin as secondary metabolites, as well as phenylalanine and ornithine as primary metabolites in leaves. However, compounds such as glutamine, glycerate, γ-aminobutyrate, and kaempferol derivatives decreased [14]. Metabolism profile in cowpea plants during drought stress indicated that compounds, such as kaempferol, amino acid derivatives, glycine, glutamine, glutamate, aspartate, asparagine, serine, alanine, threonine, sucrose, glucose, and fructose, had decreased but raffinose, catechin derivatives, unidentified phenolics, myricetin, trehalose, phenolic acids, polyols, and quercetin were induced [14].

A study of metabolomics under water stress time in the leaves and roots of cowpeas showed that leaves are more suitable than roots for recording metabolism changes [14]. It was necessary to identify if these metabolomics compounds were controlled by genetics through a distinct pathway that could help improve drought tolerance. In the leaves of both cultivars, most pathways are regulated consistently, but this is not the case in the roots [14]. Other studies show that raffinose family oligosaccharides (RFOs) [18–20] and myoinositol [21] have an important role in drought resistance. This can probably be attributed to an increase in branched-chain amino acids such as valine, leucine, and isoleucine connecting with the substrates storage for critical metabolic pathways. It is documented that drought tolerance in plants is enhanced by the overaccumulation of phenolic and other secondary metabolites [22, 23]. A study of altered metabolomics in model and forage legumes in response to drought stress by Sanchez et al. [13] showed a significant increase in organic acids, sugars, and polyols. Organic acids, including TCA cycle intermediate succinic and malic acid, along with fructose, glucose-galactose, maltose, arabitol, ononitol, and galactitol as sugars and polyols, are accumulated in response to water deficit, while glutamic, aspartic, and phosphoric acid dropped during drought [13]. The change patterns in amino acids were different. Some amino acids such as proline, leucine, and isoleucine increased, while serine, glycine, and threonine decreased. Other amino acids, such as asparagine, lysine, and valine, showed no significant changes [13]. Previous studies indicated that the accumulation of small molecules, such as compatible solutes, played an important role in drought tolerance. So, drought tolerance was enhanced in transgenic plants with a high potential for compatible solutes [24–26]. In drought stress, the observed metabolic responses may reflect the basic metabolic adaptation. Thus, conserved metabolic responses indicate not only an osmotic adjustment but also a wide range of processes including protection of membranes and proteins, radical scavenging, signaling, or buildup of reusable nitrogen and carbon repositories [27, 28].

Among the global metabolic rearrangements found in primary metabolism under water stress and after recovery from dehydration, proline increases in lotus genotypes in response to drought, reflecting the critical physiological role of proline in drought stress [13, 29]. The metabolism response of two cultivars of cowpea (Pinhel and Fradel) to drought is different. In response to drought, first primary and secondary metabolites accumulated in the root of the Pinhel cultivar, but as water stress persisted, primary metabolites decreased to roughly the control level whereas some secondary metabolites increased. In the meantime, primary and secondary metabolites initially decreased in Fradel cultivar roots, but subsequently, primary metabolites increased [14]. Using cowpea as a model species, researchers found that during drought stress plants produce organic solutes, which act as a compatible osmolyte and store them in the cytosol for the maintenance of turgor and osmotic adjustment. It is proposed that metabolite-based markers could improve drought tolerance [30]. Cowpea is a drought-tolerant legume that, during drought stress, limits water deficiency through drought avoidance strategies including stomatal closure, paraheliotropism, and moisture conservation through osmolyte accumulation [14].

There is a hypothesis that better grain yield and survival during drought and the subsequent recovery depend on organs, developmental stage, and cultivar-specific similarities in metabolic reaction. During water stress in cowpeas, nearly

88 metabolites were identified. Modifications in the metabolome profile were more intense during long periods of stress, regardless of developmental stage [14]. The most important reactions to water stress, including combinations such as quercetin 3O-6′′-malonylglycoside, kaempferol 3-O-diglycoside, quercetin, galactinol, and proline, were identified in response to water stress [14]. A study of metabolomics in the roots and leaves of cowpeas during water stress indicated that, compared to roots, leaves were more suitable for recording metabolism changes [14]. Cowpea plants respond to water stress after 12 days by stimulating the arginine/proline pathway, which may be associated with the greater damage sustained under previous drought stresses. Results showed two other pathways that have important roles: the glycine/serine/threonine pathway and the alanine/aspartate/glutamate pathway. Researchers found that the content of the glycine/serine/threonine had decreased, and using chlorophyll fluorescence data [14] faster consumption of these metabolites through enhanced photorespiration [31] is suggested. It is shown that when cowpea plants are exposed to water deficiency, they modify their metabolism to resist this condition. This mechanism occurred *via* the interplay between the shikimate and arginine/proline pathways, which induced three drought-responsive metabolites, namely proline, galactinol, and quercetin 3-O-6′′-malonylglycoside [14]. Proline, L-arginine, L-histidine, L isoleucine, allantoin, tyrosine, and tryptophan showed higher levels of accumulation in the leaves of drought-stressed plants, whereas GABA, adenosine, alanine, alpha-ketoglutaric acid, phenylalanine, choline, glucosamine, guanine, and aspartic acid decreased under drought stress [30]. Accumulation of aromatic amino acids in chickpeas during drought stress acts as a different source of energy, leading to stress tolerance in chickpeas [30].

It is found that during drought stress in the leaves of chickpeas, compounds, such as tryptophan, proline, adenosine, alanine, choline, and histidine, were highly increased, and hence, they can be defined as responses to drought conditions [30]. Metabolites, such as guanosine, guanine, phenylalanine, aspartic acid, adenosine, GABA, and choline, constantly reduced during water stress, but glutathione disulfide was induced after water stress imposition but decreased because of longer periods of water stress [30]. During drought stress, increments of amino acids, such as histidine, isoleucine, tyrosine, proline, and tryptophan at both two-time points in chickpea, occur. Previous studies showed that amino acid content increased during drought stress in soybeans [32] and *Phaseolus vulgaris* [33]. Increased levels of amino acids influenced different physiological mechanisms, including regulation of osmotic changes, disintoxication of reactive oxygen species, and adjusting the pH range in intracellular, along with enhanced stress tolerance in plants [34]. D1 protein is necessary for damage repair in photosystem II and the photosynthetic machinery generally produces various ROS. During stress conditions, the production of ROS can be induced, which can prevent D1 protein production [35, 36] and can oxidase various proteins such as D1 protein [37].

Under drought stress in chickpeas, the amount of tryptophan increased, which is involved in the ROS scavenging mechanism, leading to decreased damage to the photosystem. When chickpeas were exposed to drought stress, tyrosin and histidine as amino acids increased [30], and it is reported that tyrosin plays a role in biotic and abiotic tolerance [38–42]. Different reports indicate that arginine is an important amino acid that produces a mechanism for the transmission of nitrogen in plants during high-stress conditions [43, 44]. In chickpeas, during drought conditions, arginine content was induced [30]. In chickpeas and soybeans, during water stress, the majority of the soluble sugar and sugar derivatives were downregulated [30, 32].

It was found that the content of choline and glutathione disulfide gamma-aminobutyric in the chickpea was reduced. GABA has critical roles in plant physiology and results in tolerance under abiotic stresses *via* maintenance of cell pH, osmoregulation, and metabolism [20, 32, 45]. During water stress in chickpeas, the synthesis of some amino acids, including isoleucine, tryptophan, tyrosine, phenylalanine, and arginine, was raised [30]. The three main molecules in plant metabolism are phenylalanine, tryptophane, and tyrosine [46]. Chorismate is the precursor of these molecules that originate from the shikimate pathway. This pathway serves a basic function in the plant reproductive system, development, insect defense, and abiotic stresses [47].

3. Cold stress

Cold stress causes dehydration, resulting in plasma membrane damage. According to some studies, this might occur since the flexibility of the membrane under such conditions might be low, and it may lose its function [48, 49]. The growth and productivity of economically important crops are affected by low-temperature stress as a main abiotic stress. Osmoprotectants, such as glycine betaine, sugars (trehalose and fructans), and polyamines, were accumulated to tolerate cold stress [49]. Plants adapt to low (nonfreezing) temperatures with an accumulation of soluble sugars, among which sucrose is the most abundant [50]. Sucrose is an important signaling molecule that regulates plant metabolism under stress conditions [51, 52]. In addition, it has an important role in various processes related to the growth and development of plants [52]. Protection of cell membranes against freeze dehydration is a major factor in freezing tolerance, which is related to the accumulation of solutes such as sugars, proline, and betaines in the cytosol and alteration in membrane lipid composition [53–55]. When chickpeas were subjected to cold stress, higher levels of components such as putrescine (Put) (322%), spermidine (Spd) (45%), spermine (Spm) (69%), and the highest ratio of Put/(Spd + Spm) were observed in the tolerant genotype after the sixth day. This time, gamma-aminobutyric acid (GABA) was accumulated up to 74% more in the tolerant genotype [56]. According to reports on the effect of cold stress on the legume family, it was evident that cold stress can increase metabolites with changes in gene expression. These metabolites play a role in stress protection. Using metabolite information and genetic engineering, new cold stress-tolerant varieties of the legume crop could be developed [57].

4. Salinity stress

Salt stress removes water from the cytoplasm, causing osmotic stress. Plants respond to salt stress by the accumulation or decrease of specific secondary metabolites [58]. Polyphenol content is increased in response to biotic or abiotic stresses [59, 60]. Parida and Das showed that under salinity stress, the polyphenol content in different tissues in some plants is increased [61]. In another study, El-Shintinawy and El-Shourbagy reported that in response to salinity stress, amino acids, including cysteine, arginine, and methionine decrease, while proline concentration increases [62]. Glycine betaine is one of the nontoxic cellular osmolytes that increases osmolarity during stress periods and then helps with stress mitigation through ROS reduction, protein stabilization, and osmotic modification [63–65]. The metabolic profile in lotus legumes subjected to drought stress and salt stress indicated that their metabolic acclimation had standard features. Also, Sanchez and colleagues showed that one third to half of the metabolic changes within all lotus species were conserved [13]. Similarly, research shows that the metabolic pathways of response to environmental stress within these species are relatively conserved [66]. With increased salinity in the soil, the accumulation of ions occurs. This is one of the main abiotic stresses that negatively affect the productivity of cultivated plants. A study by Sarri and colleagues revealed that in the roots of *M. arborea* exposed to salt stress, the number of saponins, flavonoids, and triterpenic acids was reduced, while benzyl tetrahydrofurans, lignans, and phenols increased [67]. Also, research shows that the metabolic response of *vicia faba* to salt stress includes increased leaf proline and decreased fumaric and malic acid [68].

5. Heat stress

Drought and heat stresses regulate various metabolic pathways in different ways. These metabolic pathways include the metabolism of carbohydrates, amino acids, peptides, secondary metabolites, and the biosynthesis of purine and pyrimidine [17]. Heat stress in soybean decreased primary metabolic, including citrate, alpha-ketoglutarate, malate, oxaloacetate, glucose, dihydroxy acetone phosphate, succinate, pyruvate, and mannitol but drought-stress-induced carbohydrates such as glucose, fructose, sucrose, raffinose, ribose, deoxyribose, gluconate, xylose, and xylitol [17]. In a study, Liu and colleagues reported that metabolites have a critical role in response to heat stress in legume seed setting [69]. In soybean plants, the amino acid biosynthesis pathway shikimate and total aspartate family-derived amino acids are reduced in response to heat stress. So heat stress results in the reduction of tryptophan, tyrosine, phenylalanine, lysine, alanine, methionine, and isoleucine, but drought stress increases leucine, isoleucine, glycine, and proline in soybean plants. Leucine, isoleucine, proline, and glycine indicate the potential for an improved tolerance reaction under drought stress [17]. Soybean plants react to heat stress with a reduction in thymine, cytosine, and uracil as major building blocks of pyrimidine and thymidine, deoxycytidine, orotate, 2′,3′-cyclic UMP, and dUMP as a pyrimidine biosynthetic pathway, while drought stress induces the number of total metabolites in purine biosynthesis pathway [17]. Using metabolic profiling, Das and colleagues showed that upon drought stress 73% of observed flavonoids and phenylpropanoids are upregulated, and 84% of the detected secondary metabolites increase at a lower content relative to the control in reaction to heat stress, which signifies the activation of the defense mechanism for relieving stress [17]. Also, A study by Jansen, Jürgens, and Ordon shows that high temperature in *Lupinus angustifolius* increased alkaloid accumulation [70].

In soybean plants, the amino acid biosynthesis pathway shikimate and all aspartate family-derived amino acids are reduced in response to heat stress. So, heat stress results in the reduction of tryptophan, tyrosine, phenylalanine, lysine, alanine, methionine, and isoleucine [17]. Moreover, a significant reduction in phytochemicals, such as genistin, daidzin, formononetin, glycitin, syringic acid, genistein, and daidzein, is reported as the soybean plants' reaction to heat stress, but these compounds are increased in soybean plants exposed to drought stress [17]. Gill and Tuteja argued that plants alleviate stress by activating their tolerance mechanisms and antioxidant activity [71]. When soybean plants were briefly exposed to heat stress, several metabolites were downregulated. Nonetheless, under water stress, the majority of the metabolites were considerably induced or were similar to the control plants [17].

6. Flooding stress

Planting areas of Soybean, as the most important oleaginous seeds in the world, are affected by flooding. In a study, flooding stress reduced plant stand growth and grain yield and underscored the importance of genetic diversity for detecting flooding tolerance in various cultivars [72]. To distinguish the traits that mediate flooding tolerance among different soybean cultivars, Coutinho and colleagues evaluated the metabolites that play a role in flooding stress response [73]. Fumarate, succinate, pinitol, citrate, alanine, malate, and phenolic compounds are the metabolites that were detected in the leaves of soybeans upon flooding stress. Soybeans that were exposed to flooding stress had more isoflavones accumulated in the roots compared to their leaves [73]. A large percentage of the metabolites in the roots and leaves of soybeans changed by flooding stress. Browne and colleagues reported that during flooding stress, the amount of several metabolites in the leaves decreased, whereas most of the compounds in the roots increased [38]. Similarly, an important change in the primary metabolism of nitrogen and carbon was observed during flooding stress. Metabolites, such as succinate, citrate, sucrose, acetate, GABA, and alanine, accumulated in the roots, while most of these metabolites were reduced in the leaves under flooding stress [38]. Soybean plants respond to flooding stress by changing compounds that are involved in the primary metabolism of carbon and nitrogen such as carbohydrates (glucose, fructose, and sucrose), organic acids of the TCA cycle, amino acids, and metabolite profiling of isoflavone glycosides [38]. In the leaves of plants during flooding stress, the amount of fructose as a carbohydrate was decreased, but the increase in sucrose and glucose is time-dependent. This change is different in roots, where a significant increase in carbohydrates, mainly sucrose, is detected for all stress periods [38]. Previous studies showed that the amount of succinate was induced compared to the other compound of the TCA cycle [74, 75], which is in agreement with succinate accumulation and lower fumarate and malate contents in soybean plants during low oxygen conditions [73]. Soybean plants respond to flooding stress by changing compounds involved in the primary metabolism of carbon and nitrogen, such as carbohydrates (glucose, fructose, and sucrose), organic acids of the TCA cycle, amino acids, and metabolite profiling of isoflavone glycosides [73]. During flooding stress, in the leaves of plants, the amount of fructose as a carbohydrate was decreased, but the accumulation of glucose and sucrose was time-dependent. This change is different in roots, where a remarkable accumulation of carbohydrates, especially sucrose, was detected for all stress periods [73].

Previous studies showed that, compared to other compounds of the TCA cycle, the amount of succinate was induced [74, 75]. This is in agreement with succinate accumulation and lower fumarate and malate contents in soybean plants during low oxygen conditions [73]. Previous studies indicated that alanine and GABA had an important role under low oxygen conditions [76, 77]. The precursor of glycine betaine (GlyBet) is an essential osmoprotectant in plants. Research shows that GlyBet changes tolerance to drought, salinity, and other stresses [78]. During flooding stress, choline content was induced in soybean roots, while the content of this metabolite decreased in leaves. Since flooding may induce stomatal resistance and inhibit water uptake because of internally limited water [79], leaf cells need an osmoprotectant for the cellular structures. This means that the higher amount of GlyBet is part of the osmoprotectant mechanism needed in this tissue. Thus, a lower amount of choline in the leaves may show a higher demand for converting this metabolite into GlyBet [73]. It is suggested that primary and secondary metabolisms in soybean plants are strongly affected by flooding stress. Most of the changed metabolites were active in carbon and nitrogen metabolism and also in the phenylpropanoid pathway. Various responses were observed in the roots and leaves and also within flood-tolerant and flood-sensitive cultivars [73]. When *Medicago truncatula* was exposed to flooding stress, root components such as γ-aminobutyrate and alanine, accumulated raffinose, sucrose, and hexoses, while pentoses decreased. Leaves showed a significant increase in starch, sugars, sugar derivatives, and phenolics (tyrosine, tryptophan, phenylalanine, benzoate, ferulate). As a result, there was an accumulation of sugar and a reduction in organic acids in phloem sap exudates during flooding stress [73].

7. Biotic stress

Infesting alfalfa with thrips increased flavonoids and amino acids [2]. Chen and colleagues that in common beans infested with Fusarium pathogens, isoflavonoids, proline, amino acids, purines, and flavonoids were induced [80]. In addition to glycoside, terpenes, phenols, flavonoids, lipids, carbohydrates, peptides, and amino acids content were accumulated in common bean plants infected with the pathogen *R. solani* [81]. In soybean plants exposed to *Melodegyne pinodes,* and *Heterodera glycines* pressure, the number of components such as tropane, alkaloid, cysteine methionine, and phenylpropanoids increased, which could be related to the resistance properties of the crop to nematodes [30]. The same results were recorded for common beans infected with *F. solani* [80] *T. velutinum,* and *R. solani* [81]. The content of primary metabolites, including alkaloids, alcohols, and amino acids, was accumulated and Wood and colleagues found that precursor molecules of these metabolites play a role in the defense and energy supply of the plant [82]. Legumes, such as red clover, pea, and alfalfa, respond to aphid infestation by increasing triterpene, flavonoid, and saponin [83] and in alfalfa plants, infected by thrips the content of flavonoid, and amino acids were induced [84]. Another study found that soybean responds to *Melodegyne pinodes,* and *Heterodera glycines* pressure by increasing phenylpropanoids, cysteine, methionine, alkaloid, and tropane, which results in a crop resistant to nematodes [85].

8. Conclusion

It is determined that plant development and stress responses, as well as processes, such as growth, the integrity of cells, energy storing, cellular signaling, formation of membrane and scaffolding, cellular replenishing, and whole-plant resource assignment, are managed by plant metabolites [86]. There are nearly 200,000 to 1,000,000 metabolites in plants [87, 88]. In plants that produce various metabolites, changes in metabolism alter plants' physiology to adapt to various situations, such as stresses [31, 86, 89]. One of the important parts of early stress responses is changes in plant metabolism, which includes the accumulation of antioxidants to protect cellular components from oxidative damage and the accumulation of compatible solutes that retain water in the cell [15]. Phenolic compounds, such as phenolic acids, hydroxybenzoic acids, hydroxycinnamic acids, flavonoids, anthocyanins, flavonols and flavones, flavanones, and tannins, are detected in the legume family [90]. Moreover, negative and positive correlations were detected between certain metabolites and grain yield [21]. So metabolomics profiles in plants could be used as a strong selection tool to provide the correlation between phenotype and genotype and, hence, to improve plant responses to stress [30].

Polyphenol components play an important role in protecting against oxidative stresses in plants and can also indirectly protect them through the activation of endogenous defense systems and modulation of cellular signaling processes [91]. It is reported that *Medicago truncatula* responds to biotic and abiotic elicitors with changes in metabolite profiling, particularly amino acid and carbohydrate metabolism [92]. It is found that proline has an important role in stress tolerance in plants. It acts as a signaling molecule that inhibits oxidative damage [93–95]. In most stress conditions, the amount of sucrose, as a major transport sugar, was increased [96]. In addition, compatible solutes are induced during various environmental stresses. These compounds that are highly soluble in water are nontoxic. They play a role in sustaining the ordered vicinal water around proteins [26, 97]. Compatible solutes include betaines and related compounds; polyols and sugars, such as mannitol, sorbitol, and trehalose; and amino acids, such as proline [26, 98]. Previous studies indicate that compatible solutes protect plants from osmotic stress and various stress factors [26, 72, 99, 100]. Raffinose is a sugar that acts as an osmoprotectant that protects plant cells during most stressful situations, especially at the subsequent stages of stress treatment [21], and protects plants from oxidative damage [101]. GABA is another metabolite that plays a role in biotic and abiotic stress responses [40, 102, 103]. This compound protects plants against different stress situations such as protection against oxidative stress, regulation of cytosolic pH, and functions of GABA as a signaling and osmoregulation molecule [40]. Obata and Farnie showed that amino acids, including threonine, leucine, methionine, lysine, valine, and isoleucine, were usually induced during environmental stress conditions [21].

Author details

Soheila Afkar
Agriculture Department, Payame Noor University, Tehran, Iran

*Address all correspondence to: Dr.afkar@pnu.ac.ir

References

[1] Shukla PR, Skea J, Calvo Buendia E, Masson-Delmotte V, Pörtner HO, Roberts DC, et al. Intergovernmental panel on climate change. Climate change and land: Summary for policymakers.an IPCC special report on climate change, desertification, land degradation, sustainable land management, food security and greenhouse gas fluxes in terrestrial ecosystems. 2019. Chapter 4, in press

[2] Zhang Z, Chen Q, Tan Y, Shuang S, Dai R, Jiang X, et al. Combined transcriptome and metabolome analysis of alfalfa response to thrips infection. Genes. 2021;**12**(12):1-14. DOI: 10.3390/genes12121967

[3] Foyer CH, Lam H, Nguyen HT, Siddique KHM, Varshney RK, Comer TD, et al. Neglecting legumes has compromised human health and sustainable food production. Nature Plants. 2016;**2**:1-10

[4] Furlan ALF, Bianucci E, Castro S, Dietz K. Metabolic features involved in drought stress tolerance mechanisms in peanut nodules and their contribution to biological nitrogen fixation. Plant Science. 2017;**263**:12-22. DOI: 10.1016/j.plantsci.2017.06.009

[5] Singh B, Singh JP, Kaur A, Singh N. Bioactive compounds in banana and their associated health benefits–A review. Food Chemistry. 2016a;**206**:1-11. DOI: 10.1016/j.foodchem.2016.03.033

[6] Singh JP, Kaur A, Singh N, Nim L, Shevkani K, Kaur H, et al. In vitro antioxidant and antimicrobial properties of jambolan (Syzygium cumini) fruit polyphenols. LWT-Food Science and Technology. 2016b;**65**:1025-1030. DOI: 10.1016/j.lwt.2015.09.038

[7] Magalhães SC, Taveira M, Cabrita AR, Fonseca AJ, Valentão P, Andrade PB. European marketable grain legume seeds: Further insight into phenolic compounds profiles. Food Chemistry. 2017;**215**:177-184. DOI: 10.1016/j.foodchem.2016.07.152

[8] Mousavi-Derazmahalleh M, Bayer PE, Hane JK, Valliyodan B, Nguyen HT, Nelson MN, et al. Adapting legume crops to climate change using genomic approaches. Plant Cell Environment. 2019;**42**:6-19. DOI: 10.1111/pce.13203

[9] Pradhan J, Katiyar D, Hemantaranjan A. Drought mitigation strategies in pulses. The Pharma Inovvation Journal. 2019;**8**(1):567-576

[10] Muzquiz M, Varela A, Burbano C, Cuadrado C, Guillamon E, Pedrosa MM. Bioactive compounds in legumes: Pronutritive and antinutritive actions, implications for nutrition and health. Phytochemistry Reviews. 2012;**11**:227-244

[11] Zhang H, Yasmin F, Song BH. Neglected treasures in the wild-legume wild relatives in food security and human health. Current Opinion in Plant Biology. 2019;**49**:17-26. DOI: 0.1016/j.pbi.2019.04.004

[12] Handa N, Arora U, Kaur P, Kapoor D, Bhardwaj R. Role of metabolites in abiotic stress tolerance in legumes. In: Abiotic Stress and Legumes. Cambridge, Massachusetts, United States: Academic Press; 2021. pp. 245-276. DOI: 10.1016/B978-0-12-815355-0.00013-8

[13] Sanchez DH, Schwabe F, Erban A, Udvardi MK, Kopka J. Comparative metabolomics of drought acclimation in model and forage legumes. Plant,

Cell Environment. 2012;**35**(1):136-149. DOI: 10.1111/j.1365-3040.2011.02423.x

[14] Goufo P, Moutinho-Pereira M, Jorge TF, Correia CM, Oliveria MR, Rosa EAS, et al. Cowpea (*Vigna unguiculata*) metabolomics: Osmoprotection as a physiological strategy for drought stress resistance and improved yield. Frontiers in Plant Science. 2017;**8**(586):1-22. DOI: 10.3389/fpls.2017.00586

[15] Nadeem M, Li J, Yahya M, Sher A, Ma C, Wang X, et al. Research Progress and perspective on drought stress in legumes: A review. International Journal of Molecular Science. 2019;**20**(2541):1-32. DOI: 10.3390/ijms20102541

[16] Bueni PCP, Lopes NP. Metabolomics to characterize adaptive and signaling responses in legume crops under abiotic stresses. ACS Omega. 2020;**5**:1752-1763. DOI: 10.1021/acsomega.9b03668

[17] Das A, Rushton PJ, Rohila JS. Metabolomic profiling of soybean (*Glycin max*) reveals the importance of sugar and nitrogen metabolism under drought and heat stress. Plants. 2017;**6**:1-21. DOI: 10.3390/plants6020021

[18] Lugan R, Niogret MF, Leport L, Guegan JP, Larher FR, Savoure A, et al. Metabolome and water homeostasis analysis of *Thellungiella salsuginea* suggests that dehydration tolerance is a key response to osmotic stress in this halophyte. The Plant Journal. 2010;**64**:215-229. DOI: 10.1111/j.1365-313X.2010.04323.x

[19] Hochberg U, Degu A, Toubiana D, Gendler T, Nikoloski Z, Rachmilevitch S, et al. Metabolite profiling and network analysis reveal coordinated changes in grapevine water stress response. BMC Plant Biology. 2013;**13**(184):1-16

[20] Li Z, Yu J, Peng Y, Huang B. Metabolic pathways regulated by γ-aminobutyric acid (GABA) contributing to heat tolerance in creeping bentgrass (*Agrostis stolonifera*). Scientific Reports. 2016;**6**:1-16. DOI: 10.1038/srep30338

[21] Obata T, Witt S, Lisec J, Palacios-Rojas N, Florez-Sarasa I, Araus JL, et al. Metabolite profiles of maize leaves in drought, heat and combined stress field trials reveal the relationship between metabolism and grain yield. Plant Physiology. 2015;**169**(4):2665-2683. DOI: 10.1104/pp.15.01164

[22] Nakayama TJ, Rodrigues FA, Neumaier N, Marcelino-Guimarães FC, Farias JRB, CN OM, et al. Reference genes for quantitative real-time polymerase chain reaction studies in soybean plants under hypoxic conditions. Genetics and Molecular Research. 2014;**13**:860-871. DOI: 10.4238/2014. February.13.4

[23] Corso M, Vannozz A, Maza E, Vitulo N, Meggio F, Pitacco A, et al. Comprehensive transcript profiling of two grapevine rootstock genotypes contrasting in drought susceptibility links the phenylpropanoid pathway to enhanced tolerance. Journal of Experimental Botany. 2015;**66**(19):5739-5752. DOI: 10.1093/jxb/erv274

[24] Rai VK. Role of amino acids in plant responses to stresses. Biologia Plantarum. 2002;**45**:481-487

[25] Ashraf M, Foolad MR. Roles of glycine betaine and praline in improving plant abiotic stress resistance. Environmental and Experimental Botany. 2007;**59**:206-216. DOI: 10.1016/j.envexpbot.2005.12.006

[26] Chen THH, Murata N. Glycinebetaine protects plants against abiotic stress: Mechanisms and biotechnological applications. Plant,

Cell and Environment. 2011;**34**(1):1-20. DOI: 10.1111/j.1365-3040.2010.02232.x

[27] Bhatnagar-Mathur P, Valdez V, Sharma KK. Transgenic approaches for abiotic stress tolerance in plants: Retrospect and prospects. Plant Cell Reports. 2008;**27**:411-424. DOI: 10.1007/s00299-007-0474-9

[28] Hummel I, Pantin F, Sulpice R, Piques M, Rolland G, Dauzat M, et al. Arabidopsis plants acclimate to water deficit at low cost through changes of carbon usage: An integrated perspective using growth, metabolite, enzyme and gene expression analysis. Plant Physiology. 2010;**154**(1):357-372. DOI: 10.1104/pp.110.157008

[29] Diaz P, Betti M, Sanchez DH, Udvardi MK, Monza J, Marquez AJ. Deficiency in plastidic glutamine synthetase alters proline metabolism and transcriptomic response in *Lotus japonicus* under drought stress. New Phytologist. 2010;**188**(4):1001-1013. DOI: 10.1111/j.1469-8137.2010.03440.x

[30] Khan N, Bano A, Rahman MA, Rathinasabapathi B, Babar MA. UPLC-HRMS based untargeted metabolic profiling reveals changes in chickpea (*Cicer arietinum*) metabolome following long-term drought stress. Plant Cell Environment. 2019;**42**(1):115-132. DOI: 10.1111/pce.13195

[31] Kim HK, Choi YH, Verpoorte R. NMR-based plant metabolomics: Where do we stand, where do we go? Trends in Biotechnology. 2011;**29**(6):267-275. DOI: 10.1016/j.tibtech.2011.02.001

[32] Silvente S, Sobolev AP, Lara M. Metabolite adjustments in drought tolerant and sensitive soybean genotypes in response to water stress. PLoS One. 2012;**7**(6):1-11. DOI: 10.1371/journal.pone.0038554

[33] Sassi S, Aydi S, Hessini K, Gonzalez EM, Arrese-Igor C. Long-term mannitol induced osmotic stress leads to stomatal closure, carbohydrate accumulation and changes in leaf elasticity in *Phaseolus vulgaris* leaves. African Journal of Biotechnology. 2010;**9**(37):6061-6069. DOI: 10.5897//AJB09.1793

[34] Krasensky J, Jonak C. Drought, salt, and temperature stress-induced metabolic rearrangements and regulatory networks. Journal of Experimental Botany. 2012;**63**(4):1593-1608. DOI: 10.1093/jxb/err460

[35] Nishiyama Y, Yamamoto H, Allakhverdiev SI, Inaba M, Yokota A, Murata N. Oxidative stress inhibits the repair of photodamage to the photosynthetic machinery. EMBO Journal. 2001;**20**:5587-5594. DOI: 10.1093/emboj/20.20.5587

[36] Allakhverdiev SI, Murata N. Environmental stress inhibits the synthesis de novo of proteins involved in the photodamage–repair cycle of photosystem II in *Synechocystis sp*. PCC6803. Biochimica et Biophysica Acta (BBA) - Bioenergetics. 2004;**1657**:23-32. DOI: 10.1016/j.bbabio.2004.03.003

[37] Silva P, Thompson E, Bailey S, Kruse O, Mullineaux CW, Robinson C, et al. FtsH is involved in the early stages of repair of photosystem II in Synechocystis sp PCC 6803. The Plant Cell. 2003;**15**(9):2152-2216. DOI: 10.1105/tpc.012609

[38] Browne JB, Erwin TA, Juttner J, Schnurbusch T, Langridge P, Bacic A, et al. Drought responses of leaf tissues from wheat cultivars of differing drought tolerance at the metabolite level. Molecular Plant. 2012;**5**(2):418-429. DOI: 10.1093/mp/ssr114

[39] Yao LM, Wang B, Cheng LJ, WuTL. Identification of key drought stress related genes in the hyacinth bean. PLoS One. 2013;**8**(3):1-11. DOI: 10.1371/journal.pone.0058108

[40] Bouche N, Fromm H. GABA in plants: Just a metabolite? Trends in Plant Science. 2004;**9**(3):110-115. DOI: 10.1016/j.tplants.2004.01.006

[41] Mata-Pérez C, Begara-Morales JC, Chaki M, Sánchez-Calvo B, Valderrama R, Padilla MN, et al. Protein tyrosine nitration during development and abiotic stress response in plants. Frontiers in Plant Science. 2016;7:1-7. DOI: 10.3389/fpls.2016.01699

[42] Shankar A, Agrawal N, Sharma M, Pandey AK, Pandey G. Role of protein tyrosine phosphatases in plants. Current Genomics. 2015; **16**(4):224-236. DOI: 10.2174/1389202916666150424234300

[43] Flores T, Todd CD, Tovar-Mendez A, Dhanoa PK, Correa-Aragunde N, Hoyos ME, et al. Arginase-negative mutants of Arabidopsis exhibit increased nitric oxide signaling in root development. Plant Physiology. 2008;**147**(4):1936-1946. DOI: 10.1016/j.lwt.2008.10.013

[44] Brauc S, De Vooght E, Claeys M, Geuns JMC, Höfte M, Angenon G. Overexpression of arginase in *Arabidopsis thaliana* influences defence responses against *Botrytis cinerea*. Plant Biology. 2012;**14**(s1):39-45. DOI: 10.1111/j.1438-8677.2011.00520.x

[45] Barbosa JM, Singh NK, Cherry JH, Locy RD. Nitrate uptake and utilization is modulated by exogenous γ–aminobutyric acid in Arabidopsis thaliana seedlings. Plant Physiology and Biochemistry. 2010;**48**:443-450. DOI: 10.1016/j.plaphy 2010.01.020

[46] Galili G, Höfgen R. Metabolic engineering of amino acids and storage proteins in plants. Metabolic engineering. 2002;**4**(1):3-11. DOI: 10.1006/mben.2001.0203

[47] Maeda H, Dudareva N. The shikimate pathway and aromatic amino acid biosynthesis in plants. Annual Review of Plant Biology. 2012;**63**:73-105. DOI: 10.1146/annurev-arplant-042811-105439

[48] Hatfield JL, Prueger JH. Temperature extremes: Effect on plant growth and development. Weather and Climimate Extremes. 2015;**10**:4-10. DOI: 10.1016/j.wace.2015.08.001

[49] Mega S, Basu U, Kav NNV. Metabolic engineering of cold tolerance in plants. Biocatalysis and Agricultural Biotechnology. 2014;**3**(1):88-95. DOI: 10.1016/j.bcab.2013.11.007

[50] Guy GL, Huber JL, Huber SC. Sucrose phosphate synthase and sucrose accumulation at low temperature. Plant Physiology. 1992;**100**:502-508. DOI: 10.1104/pp.100.1.502

[51] Roitsch T. Source-sink regulation by sugars and stress. Current Opinion in Plant Biology. 1999;**2**(3):198-206. DOI: 10.1016/S1369-5266(99)80036-3

[52] Ho SL, Chao YC, Tong WF, Yu SM. Sugar coordinately and differentially regulates growth- and stress-related gene expression via a complex signal transduction network and multiple control mechanisms. Plant Physiology. 2001;**125**(2):877-890. DOI: 10.1104/pp.125.2.877

[53] Xin Z, Browse J. Cold comfort farm: The acclimation of plants to freezing temperatures. Plant Cell and Environment. 2000;**23**(9):893-902. DOI: 10.1046/j.1365-3040.2000.00611.x

[54] Dalmannsdóttir S, Helgadóttir A, Gudleifsson BE. Fatty acid and sugar content in white clover in relation: To frost tolerance and ice-encasement tolerance. Annals of Botany. 2001;**88**:753-759. DOI: 10.1006/anbo.2001.1465

[55] Collins RP, Helgadóttir A, Fothergill M, Rhodes I. Variation amongst survivor populations of white clover collected from sites across Europe: Growth attributes and physiological responses to low temperature. Annals of Botany. 2002;**89**:283-292. DOI: 10.1093/aob/mcf037

[56] Amini S, Maali-Amiri R, Kazemi-Shahandashti SS, Lopez-Gomez M, Sobhani-Najafabadi A, Kariman K. Effect of cold stress on polyamine metabolism and antioxidant responses in chickpea. Plant Physiology. 2021;**258-259**:1-11. DOI: 10.1016/j.jplph.2021.153387

[57] Bhat KA, Mahajan R, Pakhtoon MM, Urwat U, Bashir Z, Shah AA, et al. Low temperature stress tolerance: An insight in to the omics approaches for legume crops. Frontiers in Plant Science. 2022;**13**:1-18. DOI: 10.3389/fpls.2022.888710

[58] Mahajan S, Tuteja N. Cold, salinity and drought stresses: An overview. Archive of Biochemistry and Biophysics. 2005;**444**(2):139-158. DOI: 10.1016/j.abb.2005.10.018

[59] Muthukumarasamy M, Gupta SD, Pannerselvam R. Enhancement of peroxidase, polyphenol oxidase and superoxide dismutase activities by tridimefon in NaCl stressed *Raphanus sativus* L. Biology Plant. 2000;**43**:317-320

[60] Dixon RA, Paiva N. Stressed induced phenyl propanoid metabolism. The Plant Cell. 1995;7(7):1085-1097. DOI: 10.1105/tpc.7.7.1085

[61] Parida AK, Das AB. Salt tolerance and salinity effects on plants: A review. Ecotoxicology and Environmental Safety. 2005;**60**(3):324-349. DOI: 10.1016/j.ecoenv.2004.06.010

[62] El-Shintinawy F, El-Shourbagy MN. Alleviation of changes in protein metabolism in NaCl-stressed wheat seedlings by thiamine. Biologia Plantarum. 2001;**44**(4):541-545

[63] Gadallah MAA. Effects of proline and glycinebetaine on *Vicia faba* responses to salt stress. Biologia Plantarum. 1999;**42**(2):249-257

[64] Makela P, Karkkainen J, Somersalo S. Effect of glycinebetaine on chloroplast ultrastructure, chlorophyll and protein content, and RuBPCO activities in tomato grown under drought or salinity. Biologia Plantarum. 2000;**43**(3):471-475

[65] Saxena SC, Kaur H, Verma P, Petla BP, Andugula VR, Majee M. Osmoprotectants: Potential for crop improvement under adverse conditions. In: Plant Acclimation to Environmental Stress. New York, NY, USA: Springer; 2013. pp. 197-232

[66] Sanchez DH, Siahpoosh MR, Roessner U, Udvardi MK, Kopka J. Plant metabolomics reveals conserved and divergent metabolic responses to salinity. Physiologia Plantarum. 2008;**132**(2):209-219. DOI: 10.1111/j.1399-3054.2007.00993.x

[67] Sarri E, Termentzi A, Abranham EM, Papadopoulos GK, Baira E, Machera K, et al. Salinity stress alters the secondary metabolic profile of *M. Sativa*, *M. Arborea* and their hybrid (Alborea). International Journal of Molecular Science. 2021;**22**(4882):1-19. DOI: 10.3390/ijms22094882

[68] Richter JA, Behr JH, Erban A, Kopka J, Zorb C. Ion-dependent

metabolic response of Vicia faba to salt stress. Plant Cell Environment. 2019;**42**(1):295-309. DOI: 10.1111/pce.13386

[69] Liu Y, Li J, Zhu Y, Jones A, Rose AJ, Song Y. Heat stress in legume seed setting: Effects, causes and future prospects. Frontiers in Plant Science. 2019;**10**:1-12

[70] Jansen G, Jürgens HU, Ordon F. Effects of temperature on the alkaloid content of seeds of *Lupinus angustifolius* cultivars. Journal of Agronomy and Crop Science. 2009;**195**(3):172-177. DOI: 10.1111/j.1439-037x.2008.00356.x

[71] Gill SS, Tuteja N. Reactive oxygen species and antioxidant machinery in abiotic stress tolerance in crop plants. Plant Physiology and Biochemistry. 2010;**48**:909-930

[72] Van TTT, Hoa TTC, Hue NTN, Nguyen HT, Shannon JG, Rahman MA. Flooding tolerance of soybean (*Glycine max* L.) Merr. Germplasm from Southeast Asia under field and screen-house environments. The Open Agriculture Journal. 2010;**4**:38-46. DOI: 10.2174/1874331501004010038

[73] Coutinho ID, Henning LMM, Dopp SA, Nepomuceno AN, Moraes LAC, Gomes JM, et al. Flooding soybean metabolomics analysis reveals important primary and secondary metabolites involved in the hypoxia stress response and tolerance. Environmental and Experimental Botany. 2018;**153**:176-187. DOI: 10.1016/j.envexpbot.2018.05.018

[74] Rocha M, Licausi F, Araújo WL, Nunes-Nesi A, Sodek L, Fernie AR, et al. Glycolysis and the tricarboxylic acid cycle are linked by alanine aminotransferase during hypoxia induced by waterlogging of *Lotus japonicus*. Plant Physiology. 2010;**152**(3):1501-1513. DOI: 10.1104/pp.109.150045

[75] António C, Päpke C, Rocha M, Diab H, Limami AM, Obata T, et al. Regulation of primary metabolism in response to low oxygen availability as Reveaed by carbon and nitrogen isotope redistribution. Journal of Plant Physiology. 2016;**170**:43-56. DOI: 0.1104/pp.15.00266

[76] Miyashita Y, Dolferus R, Ismond KP, Good AG. Alanine aminotransferase catalyses the breakdown of alanine after hypoxia in *Arabidopsis thaliana*. Plant Journal. 2007;**49**:1108-1121. DOI: 10.1111/j.1365-313X.2006.03023.x

[77] Limami AM, Glévarec G, Ricoult C, Cliquet JB, Planchet E. Concerted modulation of alanine and glutamate metabolism in young Medicago truncatula seedlings under hypoxic stress. Journal of Experimental Botany. 2008;**59**(5):2325-2335. DOI: 10.1093/jxb/ern102

[78] McNeil SD, Nuccio ML, Ziemak MJ, Hanson AD. Enhanced synthesis of choline and glycine betaine in transgenic tobacco plants that overexpress phosphoethanolamine N-methyltransferase. Proceedings of the National Academy of Science. 2001;**98**(17):10001-10005. DOI: 10.1073/pnas.17122899

[79] Parent C, Capelli N, Berger A, Crèvecoeur M, Dat JF. An overview of plant responses to soil waterlogging plant stress. Global Science Books. 2008;**2**:20-27

[80] Chen L, Wu Q, He T, Lan J, Ding L, Liu T, et al. Transcriptomic and metabolomic changes triggered by *Fusarium solani* in common bean (*Phaseolus vulgaris* L.). Genes.

2020;**11**(2):1752-1763. DOI: 10.3390/genes11020177

[81] Mayo-Prieto S, Marra R, Vinale F, Rodríguez-González Á, Woo SL, Lorito M, et al. Effect of *Trichoderma velutinum* and *Rhizoctonia solani* on the metabolome of bean plants (*Phaseolus vulgaris* L.). International Journal of Molecular Siences. 2019;**20**(549):1-18. DOI: 10.3390/ijms20030549

[82] Wood C, Pilkington B, Vaidya P, Biel C, Stinchcombel J. Genetic conflict with a parasitic nematode disrupts the legumerhizobia mutualism. Evaluation Letter. 2018;**2**(3):233-245. DOI: 10.1002/evl3.51

[83] Sanchez-Arcos C, Kai M, Svatos A, Gershenzon J, Kunert G. Untargeted metabolomics appr oach reveals differences in host plant chemistry before and after infestation with different pea a phid host races. Frontiers in Plant Science. 2019;**10**:1-13. DOI: 10.3389/fpls.2019.00188

[84] Saito K, Matsuda F. Metabolomics for functional genomics, systems biology, and biotechnology. Annual Review of Plant Biology. 2010;**61**:463-489. DOI: 10.1146/annurev.arplant.043008.092035

[85] Kang W, Zhu X, Wang Y, Chen L, Duan Y. Transcriptomic and metabolomic analyses reveal that bacteria promote plant defense during infection of soybean cyst nematode in soybean. BMC Plant Biology. 2018;**18**(86):1-14

[86] Wen W, Li K, Alseekh S, Omranian N, Zhao L, Zhou Y, et al. Genetic determinants of the network of primary metabolism and their relationships to plant performance in a maize recombinant inbred line population. The Plant Cell. 2015;**27**(7):1839-1856. DOI: 10.1105/tpc.15.00208

[87] Davies HV, Shepherd LV, Stewart D, Frank T, Rohlig RM, Engel KH. Metabolome variability in crop plant species—When, where, how much and so what? Regulatory Toxicology and Pharmacology. 2010;**58**(3):S54-S61. DOI: 10.1016/j.yrtph.2010.07.004

[88] De Luca V, St. Pierre B. The cell and developmental biology of alkaloid biosynthesis. Trends in Plant Science. 2000;**5**(4):168-173. DOI: 10.1016/s1360-1385(00)01575-2

[89] Khan N, Bano A. Effects of exogenously applied salicylic acid and putrescine alone and in combination with rhizobacteria on the phytoremediation of heavy metals and chickpea growth in sandy soil. International Journal of Phytoremediation. 2017;**20**(5):405-414. DOI: 10.1080/15226514.2017.1381940

[90] Singh JP, Kaur A, Shevkani K, Singh N. Influence of jambolan (*Syzygium cumini*) and xanthan gum incorporation on the physicochemical, antioxidant and sensory properties of gluten- free eggless rice muffins. International Journal of Food Science and Technology. 2015;**5**(50):1190-1197. DOI: 10.1111/ijfs.12764

[91] Amarowicz R, Pegg RB. Legumes as a source of natural antioxidants. European Journal of Lipid Science and Technology. 2008;**110**:865-878. DOI: 10.1002/ejlt.200800114

[92] Broeckling CD, Huhman DV, Farag MA, Smith JT, May GD, Mendes P, et al. Metabolic profiling of *Medicago truncatula* cell cultures reveals the effects of biotic and abiotic elicitors on metabolism. Journal of Experimental Botany. 2005;**56**(410):323-336. DOI: 10.1093/jxb/eri058

[93] Hayat S, Hayat Q, Alyemeni MN, Wani AS, Pichtel J, Ahmad A. Role of proline under changing environments: A review. Plant Signaling and Behavior. 2012;**7**(11):1456-1466. DOI: 10.4161/psb.21949

[94] Gagné-Bourque F, Bertrand A, Claessens A, Aliferis KA, Jabaji S. Alleviation of drought stress and metabolic changes in timothy (*Phleum pratense* L.) colonized with Bacillus subtilis B26. Frontiers in Plant Science. 2016;**7**(584):1-16. DOI: 10.3389/fpls.2016.00584

[95] Cheng Z, Dong K, Ge P, Bian Y, Dong L, Deng X, et al. Identification of leaf proteins differentially accumulated between wheat cultivars distinct in their levels of drought tolerance. PLoS One. 2015;**10**(5):1-20. DOI: 10.1371/journal.pone.0125302

[96] Rolland F, Baena-Gonzalez E, Sheen J. Sugar sensing and signaling in plants: Conserved and novel mechanisms. Annual Review of Plant Biology. 2006;**57**:675-709. DOI: 10.1146/annurev.arplant.57.032905.105441

[97] Thomashow MF. Molecular basis of plant cold acclimation: Insights gained from studying the CBF cold response pathway. Plant Physiology. 2010;**154**(2):571-577. DOI: 10.1104/pp.110.161794

[98] Diamant S, Eliahu N, Rosenthal D, Goloubinoff P. Chemical chaperones regulate molecular chaperones in vitro and in cells under combined salt and heat stresses. Journal of Biological Chemistry. 2001;**276**(43):39586-395891. DOI: 10.1074/jbc.M103081200

[99] Chen THH, Murata N. Enhancement of tolerance to abiotic stress by metabolic engineering of betaines and other compatible solutes. Current Opinion in Plant Biology. 2002;**5**(3):250-257. DOI: 10.1016/S1369-5266(02)00255-8

[100] Ford KL, Cassin A, Bacic A. Quantitative proteomic analysis of wheat cultivars with differing drought stress tolerance. Frontiers in Plant Science. 2011;**2**(44):1-11. DOI: 10.3389/fpls.2011.00044

[101] Nishiyama Y, Yamamoto H, Allakhverdiev SI, Inaba M, Yokota A, Murata N, et al. Galactinol and raffinose constitute a novel function to protect plants from oxidative damage. Plant Physiology. 2008;**147**(3):1251-1263. DOI: 10.1104/pp.108.122465

[102] Shelp BJ, Bown AW, Mclean MD. Metabolism and functions of gamma-aminobutyric acid. Trends Plant Science. 1999;**4**:446-452. DOI: 10.1016/s1360-1385(99)01486-7

[103] Kinnersley AM, Turano FJ. Gamma aminobutyric acid (GABA) and plant responses to stress. Critical Review in Plant Science. 2000;**19**(6):479-509. DOI: 10.1080/07352680091139277

Chapter 8

Research on the Culture of Cabbage and the Possibilities of Increasing the Early Production

Ioana Stanciu

Abstract

The main goal of vegetable growers is to obtain early and quality harvests. In this article, the experiments were carried out in the period 2005–2007, and the culture behavior of new early cabbage hybrids, the influence of plant thickness on the production of cabbage grown in greenhouses and the field and the influence of additional protection were investigated on the early production of cabbage in greenhouses and in the field. During the years 2005–2007, studies were undertaken on the culture behavior of six hybrids and an early cabbage variety, using two different planting plots and two protection systems.

Keywords: technology, cabbage, culture, research, production

1. Introduction

The studies carried out on an international and national level attest to the importance and wide use of vegetable products, which participate more and more in human nutrition, together with meat, milk, eggs, cereals, supplementing the need for proteins, which stand out for their digestibility good, of 70–75%, and through a high quality conferred by essential amino acids [1].

The analysis of the chemical composition of vegetables shows that they contain approximately 78–93% water and 7–22% dry matter. Vegetables from the onion group have the highest dry matter content (13.5% leek, 38% garlic), and the lowest in cucumbers, lettuce, lettuce [2], and according to Bielka [3] and Maier [4], vegetables are juicy food products, with a high water content and low dry matter content; being hypocaloric foods, they decrease the intake of energy substances, preventing overweight growth.

Vegetables have a high vitamin content; the vitamins needed by the human body are taken, to the greatest extent, from vegetables and fruits. Valnet [5] states that a normal life cannot be maintained if the body does not receive the vitamins it cannot synthesize [6].

In Romania, the need for fresh vegetables, especially in the cold season, is mostly covered by imports, domestic production gains ground only as the weather warms up.

According to a study by the consulting firm Visionwise, in 2006, only 15% of all the vegetables sold on the Romanian market, through large stores, were local, while

85% were brought from Turkey, Holland, Greece, Italy and Israel. These vegetables have a bland, less pleasant taste, but excel in appearance, color and size (http://www.visionwise.ro).

Obtaining extra-early production of vegetables in our country is of particular interest, especially in the regions of the Transylvanian Plateau where, due to the cooler climate, the harvesting of vegetables begins later than in the Danube Plain or the Western Plain of the country.

White cabbage (*Brassica oleracea L., Alef. var. capitata f. alba D.C. - Brassicaceae* family) is one of the most important species in the assortment of vegetable plants in our country, a fact demonstrated by the large cultivated area that represents 20–27% of the surface total cultivated with vegetables [7].

The main goal of vegetable growers is to obtain early and quality harvests. In order to ensure high yields of early cabbage, it is necessary to study and perfect the culture technology adapted to the specific pedoclimatic conditions.

2. The situation and prospects of white cabbage culture

Vegetables are food products of vegetable origin with an important role in nutrition, due to their special sensory properties and the valuable nutrients they contain. The high consumption of fresh vegetables was determined, on the one hand, by civilization, on the other hand, by the efforts of nutrition specialists to popularize the virtues of these products.

White cabbage is an important crop in the world vegetable economy. The economic importance of the culture results from the fact that the average productions can be very high (even around 100 t/ha), the harvest is carried out over a long period of time of the year, ensuring income, practically, from spring to autumn. Cabbage is a slightly perishable product during transport, temporary storage and utilization. The expenses per surface unit are relatively low, some works, from establishment to harvesting, can be completely mechanized [8].

In 2007, green vegetables were grown on approximately 3 million ha around the world, achieving a total production of 69,214 Mt. In Europe, 496 thousand ha were cultivated, with a total production of 12,250 Mt. As shown in **Table 1**, Romania ranks first in Europe in terms of the area cultivated with cabbage, with an area of about 53 thousand ha and a total production of 1120 Mt. This places it far ahead of the following rankings: Italy (17.7 thousand ha - 345.1 Mt), Germany (14.8 thousand ha - 735.5 Mt), Spain (8.2 thousand ha - 270 Mt).

In 2005, in Cluj county, the area cultivated with cabbage was 1780 ha (**Table 1**), a relatively small area compared to the rest of the country's counties. The total production recorded in Cluj County was 40.5 Mt., of which 40.3 Mt. was realized in private properties.

Continent/Country	Harvested area (1000 ha)				
	Year				
	2003	2004	2005	2006	2007
World	3186	3079	3136	3082	3088
Africa	86	114	119	102	103
North America	149	39	38	38	38

Continent/Country	Harvested area (1000 ha)				
	Year				
	2003	2004	2005	2006	2007
South America	60	25	26	13	13
Asia	2265	2359	2410	2387	2396
Europe	250	515	516	505	496
Romania	42	42.5	54.6	45.7	53
Spain	8	7.9	7.7	8	8.2
Italy	24	17.2	18.5	17.1	17.7
Denmark	1	1.1	1.7	7.4	7.5
Austria	2	1.8	1.7	1.6	1.7
Germany	14	16	14.4	14.7	14.8
Ireland	1	0.8	0.8	0.9	0.9
Oceania	3	0.9	0.8	0.9	0.9

Table 1.
Area harvested with cabbage and other brassicas, in the world, in 2003–2007 period.

3. The importance of culture

White cabbage is grown for chefs who have a wide use in the culinary art, being used raw or pickled, with or without meat, dehydrated or frozen [9] in different ways, in an assortment of dishes much appreciated by consumers. It is also a very important raw material in the canned food industry and especially vitamin juice, usually mixed with carrot juice. Having a high resistance to low temperatures, its fresh consumption can be greatly extended in winter (4–5 months) by keeping it cold (cold storage) or in the field (in improvised shelters) in quantities necessary for consumption, from November to March. We should also not forget the preservation by pickling, a simple, effective and so frequent and traditional measure for the Romanian peasant for consumption during the winter, used as a salad or prepared in various ways, among which the sarmal cabbage or "á la Cluj" cabbage [10].

During the winter, sauerkraut is an important source of vitamins and minerals due to the fact that through lactic fermentation, they are mostly preserved without significant degradation [11]. Cabbage juice is an excellent detoxifier, being effective in gastric disorders (ulcers, gastritis, intestinal inflammation). Because it produces bloating, due to the decomposition of waste from the digestive tract, it is recommended to use it after an apple and carrot juice cure, being thus administered by oncology clinics in Germany.

Due to the high content of vitamins, cabbage is revitalizing, antiscorbutic, general rebalancing, and due to its carbohydrate and protein content, it is energetic.

4. Therapeutic value

White cabbage, apart from the fact that it is a food in demand throughout the year, is equally appreciated for its therapeutic value. Being low in fat and rich in fiber, it is used in diets, helping to keep the level of calories and fat to a minimum.

Research carried out in the last 20 years has confirmed that a regular consumption of cabbage has a beneficial effect in the prevention of colon, stomach, lung and esophagus cancer (research carried out at the University of Minnesota and J. Hopkins from the USA, in Greece, Israel, Japan, Norway).

Research has revealed that people who consume vegetables from the *Brassicaceae* family, cabbage in particular, are much less exposed than others to colon cancer (studies conducted by the American National Cancer Institute).

Sauerkraut has a high content of vitamin C and the group of B vitamins, which stimulate the body's defense system. Finnish nutritionists have recently discovered that isothiocyanates stop the evolution of cancer, and they are also maintained during lactic fermentation [12].

5. Plastics used in protected vegetable cultivation

Our country has favorable pedo-climatic conditions for the cultivation of numerous vegetable species, but the natural conditions for outdoor crops are favorable only in certain periods of the year, therefore there are interruptions in the supply of fresh products to the market. Supplementing the daily ration with a varied assortment throughout the year and not only in certain periods is of particular interest. To reduce deficit periods, vegetable growers resort to certain crop protection measures [13].

Vegetable cultivation under plastic or plastic culture, due to its specificity and particularities, is an exclusively intensive branch and cannot be conceived without the judicious use of the land in the constructions intended for the cultivation of different types of vegetables in the cold period of the year. From the data of the specialized literature, it follows that crop successions ensure an earlier appearance of production in the spring and an extension of it in the autumn [14].

The plastics used in vegetable cultivation are of various types and present multiple improvements. The first experiments regarding the use of plastic masses, in the culture of vegetables, were made in 1929, at the University of Kentucky - USA. The experiments were carried out using cellophane, under which lettuce was grown. In 1932, at the Institute of Agricultural Physics in Leningrad-Feinberg, research was initiated with cellophane shelters on a radish crop. The experiments were resumed in 1948, but with a wider application in vegetable cultivation: 1950 in Japan, 1954 in Russia and the USA, 1956 in England and France.

6. Other materials used in direct crop cover

Advances in science have allowed the improvement of culture technologies for some vegetable plants, with high yield methods being developed [15]. Thus, the use of non-woven textiles in horticulture occupies an important role in modern technologies.

Non-woven textiles are part of the group of materials used in horticulture, generically called agrotextiles. These materials are made by a specific technology from polymer fibers of a polyolefinic nature, such as polypropylene, polyesters or polyamides. In horticulture, these materials are used in various forms, such as mulch and as covering material in the blanket system.

Polypropylene was first introduced in 1957 by an Italian company Montecatini (chemical compound) and proved to be a cheaper material than polyethylene. It has a

Figure 1.
Agryl P-19 coating material.

thickness of 10–30 microns, representing a very high transparency, very good dimensional stability, mechanical and chemical resistance, against oils, acids, bases.

Non-woven textiles are presented as a "fabric" in which the constituent fibers are not woven but are overlapped, by bonding, in a layer of variable thickness (0.1–0.2 mm). It is a heat-resistant polymer used in the pre-packaging of vegetable products, making textiles, ropes, reusable containers.

The manufacturing technology of these materials provides for the random arrangement of fibers in the same plane in a layer uniform in thickness and composition, followed by pressing and thermoadhesive bonding between two rolls, a process called calendering [16].

AGRYL P19 (**Figure 1**) - is a light and resistant material, has a weight of 19 g/m^2, protects against bad weather (cold, wind) and insects, improves yield and quality of early production; creates a favorable microclimate for plant growth; it has a light flux transmissivity capacity of 85% [17–21].

7. Solar culture technology

The experimental culture was established in the greenhouse covered with double polyethylene film.

The land was prepared in the fall, it was fertilized with 40 t/ha of manure, which was incorporated with the motor cultivator. In early spring (17.03.2005, 15.03.2006, 07.03.2007) the solarium was covered with durable polyethylene film TVK.

The land was modeled in layers of 140 cm. The seedlings were produced in the heated greenhouse, the sowing was carried out on 14.01.2005, 11.01.2006 and 10.01.2007 respectively. On 28.01.2005, 30.01.2006 and 26.01.2007 the seedlings were transplanted into nutrient cubes of 5×5 cm.

The seedlings were planted on 24.03.2005, 20.03.2006 and 9.03.2007, when temperature 8°C were recorded in the soil of the greenhouse. The planting distances were as follows: 40 cm between rows and 35 cm between plants per row, thus resulting in a plot of 71,500 plants/ha, respectively 40 cm between rows and 40 cm between plants per row, achieving a plot of 62,500 plants/ha. Immediately after planting, the double-protected varieties were coated with Agryl P-19.

In the experimental year 2005, 1 week after planting, the gaps were filled with seedlings from the same hybrid, then on 04/03/2005, facial fertilization was done with Complex III 250 kg/ha. On 16.05.2005, another foliar fertilization with Fertitell 1.5 l/ha was carried out.

In the experimental year 2006, fertilization with Complex III was carried out on 29.03, and in the experimental year 2007, fertilization was carried out on 21.03.2007.

Thirteen waterings were carried out in 2005, 11 in 2006, with quantities of 200–250 m^3 of water. In 2007, six waterings were carried out with quantities of 200–250 m^3 of water, and the rest of the waterings were done with the help of a drip irrigation system, with a flow rate of 2 l/hour, through intermittent applications.

During the vegetation period, three manual weedings were carried out in the period 12–28.04.2005, two in the period 04–20.04.2006 and only one manual weeding in the period 22.03–06.04.2007.

I mention that no phytosanitary treatment was done during the 3 years of experience.

The removal of the covering material, Agryl P-19 type, was done according to the climatic conditions in the place of cultivation, as follows: on 16.05.2005, 20.04.2006 and 23.04.2007. This material was kept on culture for 50 - 30 - 25 days, respectively.

The harvests were carried out between 20.05-10.06.2005, when productions varying between 62.8–75.5 t/ha were obtained, in 2006 the harvests were made between 15.05-5.06.2006 with productions between 59.1–78.0 t/ha, and in 2007, the harvesting period was between 03.05-30.05.2007, obtaining productions between 61.2–76.5 t/ha.

8. Field culture technology

The preparation of the land started already in the autumn, when it was fertilized with 25 t/ha of manure, which was incorporated with deep plowing, at 28–30 cm.

The land was modeled in layers of 140 cm. The seedlings were produced in the heated greenhouse, the sowing was carried out on 14.01.2005, 11.01.2006 and 10.01.2007 respectively. On 28.01.2005, 30.01.2006 and 26.01.2007 the seedlings were transplanted into nutrient cubes of 5 × 5 cm. The seedlings were planted on 29.03.2005, 22.03.2006, respectively on 20.03.2007 when the soil temperature reached temperature 8°C. The planting distances were as follows: 40 cm between rows and 35 cm between plants per row, thus resulting in a plot of 71,500 plants/ha, respectively 40 cm between rows and 40 cm between plants per row, making a plot of 62,500 plants/ha. Immediately after planting, the protected varieties were coated with Agryl P-19.

The covering material, type Agryl P-19, was removed, following the monitoring of the climatic conditions in the field, as follows: on 23.05.2005, 17.05.2006 and 14.05.2007. This material was kept on culture for 55 - 40 - 30 days, respectively.

In the experimental years 2005–2006, 1 week after planting, the gaps were filled with seedlings from the same hybrid and facial fertilization was done with Complex III 250 kg/ha. In 2007, after filling in the gaps, two partial fertilizations were made, one on 29.03, with 150 kg/ha of ammonium nitrate and on 20.04 with the same amount of mineral fertilizer.

Three waterings were carried out in 2005, seven waterings in 2006 and eight waterings in 2007, with quantities of 200–250 m^3 of water. I mention that in 2005 there was a rainy period, for this reason no more irrigation was done until the harvest.

During the vegetation period, three manual weedings were made between 08.04–10.05.2005, two between 01.04–20.05.2006 and two between 30.03–02.05.2007. During the vegetation period, two phytosanitary treatments were carried out, one for *Delia radicum* L. with Fastac 0.1 l/ha and one treatment with Optimol 15 kg/ha for *Deroceras agreste* L. The cabbage fly attack depreciated part of the harvest unprotected, therefore the production of early cabbage from the field was lower in 2005 than in the following years of experience.

The harvests were carried out in the period 06.06–14.06.2005, when productions varying between 38.4–59.1 t/ha were obtained, in 2006 the harvests were made in the period 06.8–19.06.2006 with productions between 48.3–58.4 t/ha, and in 2007 productions varying between 53.4–61.7 t/ha were obtained, harvesting was carried out between 25.05–04.06.2007.

9. Conclusions

Covering the early cabbage culture with textile tarps greatly influences the microclimate of the culture space. The temperatures in the greenhouse and field have the highest values, in the first 10 days after planting, under the covering material. Thus, in the solar the average air temperature is higher by temperature 1.75°C at 8 o'clock and by 2.25°C at 12 o'clock, under the covering material, and in the field, the difference between the air temperature in the open and that under Agryl, shows an average of 2.15°C at 8 o'clock and 2.40°C at 12 o'clock.

Covering the early cabbage culture with agrotextiles also positively influenced the evolution of soil temperatures, with average differences of 2.15°C between the covered soil at 8 o'clock and 2.50°C at 12 o'clock, in the solar culture and of 1.37°C at 8 o'clock and 1.70°C at 12 o'clock, for field crops.

Analyzing the effect of the protective material on the atmospheric humidity, it is found that this parameter showed higher values, both in the solar and in the field. Thus, the average difference, between the two methods of protection, in terms of relative atmospheric humidity is 16.2%, at 8^{oo} and 24.3% at 12 o'clock, and for field crops the average difference is 11.0%, was registered at 8^{oo} and 17.4% at 12^{oo}.

The humidity of the soil, from the solar, evolved as follows: at a depth between 0 and 10 cm, on average, the humidity was higher by 7.82%, in the case of Agryl coating, and at a depth between 10 and 20 cm, the average difference in humidity was 3.44%, compared to the uncovered crop. In the field culture, the average difference in soil moisture was 5.1%, at a depth between 0 and 10 cm and 8.3% at a depth of 10–20 cm, in favor of the covered culture and was preserved more constant during the vegetation.

Following the observations made on the light intensity, both in the experimental culture in the greenhouse and in the field, it is highlighted that the variants covered with Agryl P-19 benefit from a lower amount of light than the uncovered ones, so that in on average, there is a daily difference of 1544 lux, at 8^{oo} and of 2536 lux, at 12^{oo}. With the early cabbage crop in the field, an average daily difference of 1347 lux is recorded, at 8^{oo}, and at 12^{oo}, between the two culture systems, an average difference of 2016 lux is found, in favor of the plants uncovered.

- Covering the early cabbage crop from planting with agrotextile tarps has favorable effects on the growth and development of the plants, so that all the morphological characters (rosette diameter, plant height, number of leaves in the rosette, head weight, leaf weight) were positively influenced and they recorded higher values, both in the greenhouse and in the field, compared to the uncovered crops. Biometric observations were made 19–28 days after planting.

In the culture established in the greenhouse, the diameter of the rosette of the leaves registers average values of 43.6 cm under the influence of the Agryl coating and only 36.6 cm, in the uncovered plants. In field crops, direct coverage increases rosette diameter to 30.4 cm compared to only 23.2 cm in the control.

The length of the stem registers a slight elongation in the covered crops, presenting an average waist of 7.4 cm, compared to 5.93 cm, in the control. This elongation is due to the lack of light, in the plants grown in the space covered with Agryl. In the field, the stem length is shorter than in the greenhouse; in those protected with Agryl, this character had an average of 4.25 cm, and in the non-coated versions an average of 3.65 cm.

Covering the plants with Agryl influences to a lesser extent the number of leaves in the rosette, in plants grown in the solar system. In the covered variants, the plants have an average of 11.9 leaves, and in the uncovered variants, a number of 10.7. In the field, the differences are greater than 13.5 leaves in the rosette, and in the uncovered variants 10.5 leaves.

Covering the early cabbage crop with Agryl has less influence on the average weight of the heads, in solar being 0.98 kg, for the covered versions and 1.02 kg, for the uncovered ones. In field culture, the average weight of the heads register values very close to 0.80 kg, for the protected variants and 0.75 kg, for the unprotected variants.

The total weight of the plant and the outer leaves are much higher in the culture covered with Agryl and reaches 1.94 kg, respectively 0.99 kg, and in the simply protected variants 1.84 kg, respectively 0.79 kg, in the culture from solar. For field crops, the values are lower than 1.46 kg, respectively 0.39 kg for the covered variants and 1.38 kg, respectively 0.38 kg, for the uncovered variants.

32.6 leaves in the head were determined, for the double protected variants and 30.2 leaves, for the single protected variants.

The elongation of the inner coccine, in the covered plants, is due to the lack of light, under the covering tarpaulin, which prints a slight tendency of the stem elongation during the growth period and registers the following average values: 7.78 cm, in the double protected variants and 6.53 cm, for the simply protected variants from the solar and 5.70 cm, respectively 5.12 cm for the culture in the field.

- The degree of weeding of the cabbage crops, from the two variants of cabbage culture, both in the greenhouse and in the field, is reduced when the Agryl type cover material is used. In solar, protection reduced the number of weeds by about 50%, and in field culture, direct covering caused a lower degree of weeding by about 23.8%.

- The dynamics of early cabbage harvests in the greenhouse show that the best start of the harvests, in 2005, was recorded in the third decade of May, when a 71,500 pl./ha crop was used, with protection simple. In 2006, the earliest harvests were obtained in the second decade of May, at the rate of 62,500 pl./ha, and in 2007, benefiting from favorable climatic conditions, the harvests began in the first decade of May, for the variants with a plot of 71,500 pl./ha.

The dynamics of harvests in the field crop show that the highest productions, at the first harvests, are recorded in the variants covered with Agryl, cultivated at a plot of 71,500 pl./ha. For crops not covered with textile tarps, harvesting is delayed by 4–7 days compared to covered crops.

- Regarding the influence of the hybrid on the extra-early production, it is found that the highest productions were registered with the Surprise and Santorino hybrids, which obtained average productions of 39.63 t/ha and 39.29 t/ha, respectively.

Very good extra-season products are obtained if a planting plot of 71,500 pl./ha is used. This planting plot determined, on average, extra-early productions of 39.29 t/ha, with a very significant increase of 13.8% compared to the crop with a smaller plot.

Covering with Agryl P-19 shows its effect in the years 2006–2007, when this technological measure to increase earliness generates average extra-early productions of 35.73 t/ha with a very significant difference of 2.38 t/ha, compared to simply protected variants. These results can also be attributed to the fact that in these years the crop remained covered for a period of only 30 days from the beginning of vegetation.

- The highest total productions, per surface unit, in 2005, were obtained with the Resistor and Parel hybrids, and in 2006–2007, with very close values, with the Santorino hybrids (70.68 t/ha) and Surprise (69.03 t/ha), higher productions at a plot of 71,500 pl./ha.

The method of additional protection with Agryl tarpaulins of cabbage in the greenhouse increases the total production on average by 4.7%, determining a very significant difference in production of 3.13 t/ha, compared to the simply protected variants.

In field crops, the most productive were the hybrids Santorino (56.99 t/ha) and Surprise (56.78 t/ha), at a plot of 71,500 pl./ha, in the variants covered with agrotextiles, a period of 30–40 days. 7. The researched factors influence to a lesser or greater extent the biochemical composition of the heads of cabbage.

Thus, in crops covered with Agryl, the content of soluble dry matter is lower due to reduced light conditions, especially in solariums where it is filtered through two layers of coating.

The content of vitamin C is influenced in the same way, the highest values of 31.68 mg/100 g of fresh substance being registered in the field culture, in the version not covered with Agryl, and the lowest, in the solar culture, in the plants coated with Agryl (21.2 mg/100 g s.p.).

The Santorino hybrid stands out for its higher content of dry matter and vitamin C, compared to the other cultivars.

In general, all the biochemical characters, in the cabbage grown in the field, showed higher values than in the cabbage grown in the greenhouse, and the nitrate content was lower, compared to the heads analyzed in the greenhouse. In protected spaces, due to the conditions of lower light, cabbage accumulates higher amounts of nitrates (110.7 mg/kg), values that are found at the lower limit of the content specified in the specialized literature (30–580 mg/kg) and much below the maximum limit allowed for this food product.

Author details

Ioana Stanciu
Faculty of Chemistry, Department of Physical Chemistry, University of Bucharest, Bucharest, Romania

*Address all correspondence to: istanciu75@yahoo.com

References

[1] Butnariu H, SI Collab. Legumicultură. Bucuresti: EDP; 1992

[2] Lamont WJ. Plastics: Modifying the microclimate for the production of vegetable crops. HortTechnology. 2005;**15**:477-481

[3] Mochizuki MJ, Rangarajan A, Bellinder RR, Bjorkman T, Van Es HM. Overcoming compaction limitations on cabbage growth and yield in the transition to reduce tillage, HortScience. 2007;**42**(7):1690-1694

[4] Maier I. Cultura Legumelor. Bucuresti: Agro-Silvică; 1969

[5] Valnet J. In: Ceres, editor. Tratamentul bolilor prin legume, fructe si cereale. Bucuresti; 1986

[6] Hoza G. Sfaturi practice pentru cultura legumelor. Bucuresti: Ed. Nemira; 2003

[7] Posta Gh, Berar V. Studies of the analysis of morphological characters from a range of late cabbage hybrids cultivated in the field conditions. Bulletin UASMV Cluj-Napoca, Horticulture. 2208;**65**(1):153-156

[8] Dumitrescu M, SI Collab. Producerea Legumelor. Bucuresti: Artprint; 1998

[9] Chilom P, Balasea M, Dinu M, Ghe P, Spirescu C. Îngrăsământ biologic complex cu stimulatori de crestere pentru fertilizarea foliară, Revista Sănătatea Plantelor, nr. 6-9, Bucuresti. 2000

[10] Glaman, Gh., Popa N. Comportarea unor soiuri/hibrizi de varză albă în conditiile pedoclimatice de la ferma Ciolpani-Ilfov. Revista Hortinform, nr. 2/150. 2005

[11] Apahidean ALS, Apahidean M, Maniutiu D, Ganea R, Paven I, Ficior D, et al. Influenta metodei de protejare la varza cultivată în solar, Buletinul USAMV-CN nr. 61/2004. 2004

[12] Iburg A. Remedii Naturiste. Bucuresti: Allfa; 2006. pp. 262-262

[13] Ciuciuc E, Balasa M. Cercetări privind stabilirea unor metode pentru obtinerea de productii extratimpurii la pepene galben cultivat pe soluri nisipoase. Revista Hortinform, nr:10/86, Bucuresti. 1999

[14] Novika RGH, Hutomo CS, Wahidah NJ, Sumarno L, Rahmawati NY, Ansori ANM, et al. The effect of Apium graveolens L. in progesterone-induced blocking factor (PIBF) during pregnancy. Research Journal of Pharmacy and Technology. 2022;**15**(10):4463-4468

[15] Posta Gh, Berar V. Researches concerning the yield performances of the early cabbage hybrids cultivated in the field conditions. Buletinul USAMV ClujNapoca, Horticultura. 2007;**64**(1-2):113-117

[16] Masamba KG, Nguyen M. Determination and comparation of vitamin C, calcium and potassium in four selected conventionally and organically grown fruits and vegetables. African Journal of Biotechnology. 2008;**7**(16):2915-2919

[17] Waterer D. Demonstration of Multirow Floating Covers for Vegetable Crops. Canada: University of Saskatchewan; 1992

[18] Gerszberg A. Tissue culture and genetic transformation of cabbage (Brassica oleracea var. capitata): An overview. Planta. 2018;**248**:1037-1048

[19] Joraboevich SA, Sanakulovich LS. Selection of promising varieties of white cabbage for cultivation in re-culture. Pioneer: Journal of Advanced Research and Scientific Progress. 2022;**1**(4):144-150

[20] Zhao D et al. Impact of lactobacillus paracasei HD1. 7 as a starter culture on characteristics of fermented Chinese cabbage (Brassica rapa var. pekinensis). Food Science and Technology Research. 2016;**22**(3):325-330

[21] Siciliano I et al. Mycotoxin production in liquid culture and on plants infected with Alternaria spp. isolated from rocket and cabbage. Toxins. 2015;7(3):743-754

Chapter 9

Abiotic Stress-Tolerant Crop Varieties in India: Status and a Way Forward

Boraiah K.M., Basavaraj P.S., Vijaysinha D. Kakade, Harisha C.B., Pratapsingh Khapte, Halagundegowda G.R., Krishnamurthy D., Neeraj Kulshreshtha, Vijayakumar H.P., Bhojaraj Naik, Jagadish Rane Sammi Reddy K. and Himanshu Pathak

Abstract

The abiotic stresses, such as drought, waterlogging, heat, cold, and salinity, cause significant crop yield losses associated with extremes of moisture and temperature and ion imbalance. The occurrence of these conditions is being aggravated by climate change, global warming, and industrial pollution. It is crucial to safeguard food security through a constant and sustainable crop production system under multiple abiotic stresses. The cultivation of climate-resilient varieties is one of the best strategies being followed across diverse agroecosystems in the world including India to mitigate the impact of abiotic stress on crop production. Indian agricultural institutional network under the umbrella of the National Agriculture Research System developed a good number of abiotic stress-tolerant varieties across the field and horticultural crops. However, only a few crops' varieties' introgression with *SUB1* gene and salinity tolerant QTLs are being cultivated largely in the areas prone to submergence and salinity stress, respectively. In this book chapter, we have updated the status of abiotic stress tolerance crop varieties (ASTCVs) along with stress-wise trend analysis to disseminate information among farmers, students, scientists, and policymakers involved in abiotic stress management. Finally, we also discussed the strategies to reorient the breeding program to develop climate-smart varieties with multiple biotic and abiotic stress tolerance.

Keywords: climate change, abiotic stress, drought, waterlogging, heat, salinity, genomics, phenomics, climate-smart varieties

1. Introduction

Agriculture is a livelihood for about 60% of India's population, and it is significantly contributing to the national economy. It has been proved even during covid pandemic situation. Although other sectors of the economy dominating in the gross domestic product (GDP) with faster growth rates and agriculture contribution reduced to less than 20%, agricultural production has enhanced substantially. With the advent of new innovations and technologies in agriculture since after independence, India translated from the begging bowl to being self-sufficient in food production, and even emerged as a net exporter of agriculture and allied products. The food grain production increased over six times to over 314 Mt. in 2022 from 51 million tonns (Mt) in 1950–1951 [1]. On the other hand, increase in the demand for quantity, quality, and nutritious and diverse food due to the increasing population, average income, and awareness on health and globalization. Fulfilling this demand is a serious challenge as the nation often witnesses frequent droughts, floods, heat, and cold waves, as well as incidences of pest and disease epidemics resulting in losses in agricultural productivity [2–4]. Further, arable soils become saline, sodic, and acidic as a consequence of frequent floods, high temperatures, indiscriminate use of chemical fertilizers, and also due to the use of polluted or sewage water for irrigation. These conditions are further worsened due to changing climate, which poses a serious threat to food security in the future. Although the negative impacts of climate change-driven abiotic stresses are global, countries such as India are more vulnerable in view of the greater proportion of its population depending on rainfed-based agriculture with limited natural resources.

In this book chapter, we have briefly discussed the impact of abiotic stresses on crop husbandry and enlisted percent yield losses that occurred in different crops due to drought as a shred of evidence. Further, mentioned the different management strategies to cope up with the abiotic stresses and discussed about climate-resilient varieties developed by associate institutes of the National Agricultural Research System (NARS) for different abiotic stresses with a brief trend analysis and status on varietal development for different abiotic stress tolerance and their adoption with citing success stories. Finally, discussed strategies to upscale from the development of climate-resilient varieties to climate-smart crop varieties as a way forward.

2. Impact of abiotic stresses on crop production

Abiotic stresses affect agricultural production and its productivity and cause losses up to 50% of overall agricultural production depending on their intensity. The loss in crop production alone is projected 10–40% in India by 2080–2100 unless we adapt and mitigate global warming (IPCC 2007). For instance, a rise in temperature by 1.5°C and a reduction in the precipitation of 2 mm can reduce the yield of rice by 3 to 15% [5]. Similarly, the loss in wheat production in Indo-Gangetic plains would be 4 to 5 million tonnes with 1°C rise in temperature [6]. The National Institute of Abiotic Stress Management located in Pune Maharashtra compiled information on the impact of different abiotic stresses on agricultural crops along with management strategies to prevent yield loss [1, 7]. The prevalence of high temperatures and often unforeseen hailstorm events coupled with unseasonal heavy rainfall in the central and southern parts of the country were increasing in recent years. These events cause sunburn and fruit cracking in fruit crops particularly berries and other temperate fruit crops [8, 9]. Similarly, extreme temperature and moisture coupled with high or low relative humidity adversely affect productivity and quality beside aggravating the pest and diseases on vegetable crops.

3. Strategies for coping with abiotic stresses in Indian agriculture

Abiotic stresses together with biotic stresses are threatening crop production and consequently food and nutritional security. Further, the situation is likely to worsen with projected climate change. For combating the effects and impact of the resultant abiotic stressors, various approaches can be employed by adopting advanced science and policy initiatives including government schemes such as Pradhan Mantri Krishi Sinchai Yojana (PMKSY) to save water and expand irrigation facilities and Pradhan Mantri Fasal Bima Yojana to provide insurance against crop loss due to unforeseen extreme natural phenomena. Improved agricultural practices such as water-saving technologies such as *in situ* and *ex situ* moisture conservation, water harvesting for supplemental irrigation, residue incorporation, developing and adopting/cultivating tolerant crop varieties, conservation agriculture, site-specific nutrient management practices, etc. are some of the strategies to cope up with abiotic stresses. Adoption and cultivation of varieties with tolerance to the deficit and excess moisture, heat, cold, and salinity in vulnerable agroecosystems play a crucial role. This relies on the development, popularization, and availability of abiotic stress-tolerant crop varieties (ASTCVs) to farming communities.

Plant science experts of the Indian NARS including various institutes of the Indian Council of Agricultural Research (ICAR) and State Agricultural Universities (SAUs) are making concerted efforts over the years for developing improved varieties with enhanced tolerance to different abiotic stresses. Some of these climate-resilient crop varieties are being cultivated by farming communities in different parts of the country in the event of extreme weather situations to mitigate the adverse impact of abiotic stresses on crop production. The adoption and cultivation of such varieties by the farming community across the world including India contributed to sustainable food security even under changing climate scenarios [10–13]. However, the lack of information on abiotic stress-tolerant crop varieties in one place is a prime limiting factor for awareness, adoption, and cultivation of such varieties on a large scale. Only a few organizations working on abiotic stresses including ICAR-NIASM and ICAR-CRIDA attempted to gather such data for documenting and disseminating the information on abiotic stress-tolerant crop varieties [14]. In this book chapter, we updated the comprehensive information on abiotic stress-tolerant crop varieties along with a brief and critical analysis of stress-wise trends in the number of ASTCVs and also cited some success stories of the adoption of climate-resilient varieties. Further, we also discussed the strategies to develop and assess climate-smart crops for mitigating the adverse effects of climate change on crop production.

4. Climate-resilient varieties for abiotic stresses

Enhanced and sustainable crop production is key to stabilizing and doubling the farmer's income by minimizing the risk of prevailing abiotic stresses in agriculture in order to retain the farming community in the agriculture sector and improve their livelihoods. This also ensures food and nutritional security even under abiotic stress situations driven by aberrant weather conditions. Many policymakers and experts worldwide from the public and private sectors have suggested that the development and identification of climate-resilient varieties with enhanced tolerance to different abiotic stresses are crucial to sustaining and improving crop yields under changing climate scenarios. In this regard, the Indian National Agricultural Research System

(NARS) including various ICAR institutes and State Agricultural Universities are making concerted efforts over the years for developing climate-resilient varieties of different crops with enhanced tolerance to different abiotic stresses. For instance, ICAR-NRRI, Cuttack, Odisha developed and released rice varieties tolerant to submergence and flooding, and high-temperature conditions. Similarly, ICAR-CSSRI, Karnal developed salinity-tolerant varieties in rice, wheat, mustards, etc. Beside, the intervention of the private campiness through R&D in the identification of genetic resources and development of varieties and hybrids particularly in vegetable crops gained momentum in the past decade.

The data on ASTCVs developed by NARS and their relevant information collected from diverse sources such as agricultural public institutional (ICAR and SAUs) publications, websites, seed portals, research and review papers, and documents from national and state government was scrutinized, compiled, and used for trend analysis. It was found that about 750 crop varieties (**Figure 1A**) were reported to be tolerant for different abiotic stresses such as drought (65%), flood/waterlogging (5%), high temperature (17%), low temperature (5%), and salinity/acidity (8%). It indicates drought followed by heat stress is more prevalent and frequently affects the agri-based food production system though other stresses affect often it. Further, data also gave insights into crop-wise abiotic stress-tolerant varieties. More number varieties belong to field crops including cereals, pulses, and oilseeds indicating the importance of crops and crop priority in the abiotic stress resistance breeding program. Trends on stress-wise tolerant crop varieties are discussed in the next section along with traits/mechanisms explored in abiotic stress tolerance breeding program and cited some of the successful examples.

4.1 Drought-tolerant crop varieties

Drought is one of the serious threats to crop production in most parts of India, particularly in arid and semiarid regions. Water deficit affects the number of plant growth and developmental process. In turn, plants have different responses at morphological, biochemical, and physiological levels to deficit moisture conditions

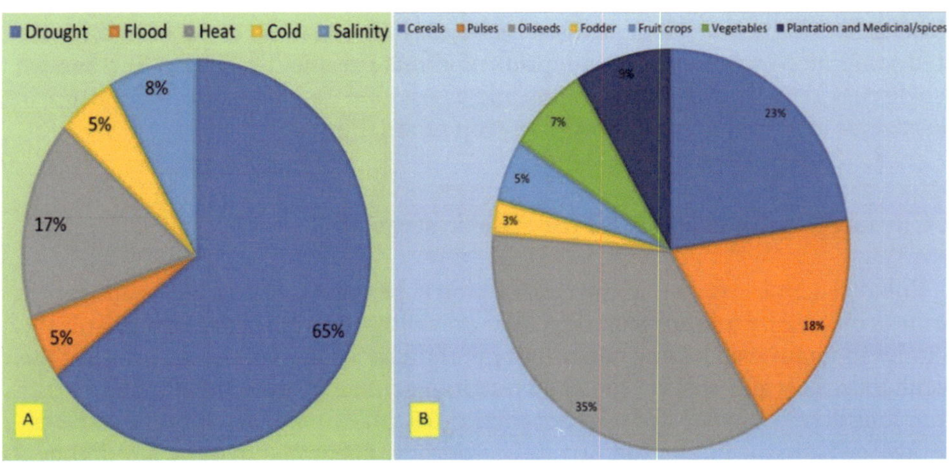

Figure 1.
Proportion of agricultural crop varieties tolerant to different abiotic stresses (A) and crop group wise distribution of drought tolerant varieties (B).

to adapt and survive. A number of adaptive traits have been studied and used for the improvement of drought tolerance such as early vigor, early flowering, osmotic adjustment, leaf senescence, stay green, etc. Plant breeders of several institutes across the world including ICAR and state agricultural universities in India explored the traits associated with early flowering, better root architecture, and efficient photosynthesis [7] to develop drought-tolerant varieties.

In India, among the reported ASTCVs, more than 65% of crop varieties are drought tolerant. Among drought-tolerant varieties, more number varieties belong to field crops (**Figure 1B**) including cereals (23%), pulses (18%), and oilseeds (35%) indicating the scope and priority for the drought resistance breeding program. The varieties of rice, wheat, sorghum, and pearl millet are shared a major proportion of drought-tolerant cereal crops. Though this may not be representing the holistic crop component since millets, which are naturally and inherently multiple stress tolerant, including drought were not covered due lack of reliable information. However, this trend definitely indicates some realistic trend of major and staple food crops and their importance in drought tolerance breeding. A similar trend has been observed in pulses also where major pulse crops such as pigeon pea and chickpea dominate among drought-tolerant pulse crop varieties. Whereas, among the oilseeds though most of them cultivated in rainfed condition except hybrid-based sunflower and mustard, reported a greater number of drought-tolerant varieties in groundnut. This is mainly due to the large cultivated area of groundnut frequently witnessing the drought at critical growth stages beside its unique reproductive stages, such as peg development and penetration, were sensitive to deficit moisture. This indicates potential scope for the development of drought-tolerant varieties in groundnut with special emphasis on maximizing pod-peg and mature-immature pod ratio. Beside field crops, there are number of fruits and vegetable crops reported tolerant to drought (**Table 1**).

There are several pieces of evidence on successful breeding efforts in the development of drought-tolerant varieties [15, 16] and subsequently, case studies reported the impacts of such varieties [17] in reducing the risk of drought and sustaining the livelihood of farmers. Further, under the NICRA project, it was demonstrated that short duration and drought-tolerant rice varieties such as Sahbhagi dhan, Anjali, Naveen, Birsa Vikas Dhan 109′ (85 days duration), and 'Abhishek' withstand up to 2 weeks of exposure to dry spells in northern and eastern states. Such varieties are the best choice for drought-proofing in rainfed rice cultivation as they provide a significant yield advantage in drought years over the traditional long-duration varieties. Similarly, in groundnut, several varieties including some recently released varieties, such as Dharani, Co7, Ajeya, Vijetha, DH-257, Dh 256, and Kadiri Lepakshi, were reported as drought tolerant (**Table 1**) and a few of them particularly, Kadiri Lepakshi gaining wide popularity due to its high yielding and resistance to pest and diseases. However, farmers are being reported problems of irregular maturity and immature pods with poor quality kernels in rainfed areas, which most of the time witness drought situations due to the gambling of monsoon winds. Hence, it is essential to consider the quality and other traits during the abiotic resistance breeding program.

4.2 Waterlogging or submergence-tolerant crop varieties

In recent years, Indian agriculture witnessing frequent occurrences of waterlogging conditions due to increasing events of heavy rainfall. For instance, currently, about 20 million hectares of the world's rice-growing area are at risk of occasionally being flooded to submergence level, particularly in major rice-producing countries such as

Crop	Varieties
Field crops	
Rice	Kalinga 1& 2, Prabhat, Vandana, Anjali, Naveen, Abhishek, Shabhagidhan, Kamesh, Shusk Samrat, Satyabhama, CR Dhan 40, 200, 201, 202, 203, 204, 205, 206, 207, 209, 210, 801, Purna, Sahbhagi, CAUR-1, MTU-1010, Pant Dhan 16, Barani Deep, Indira Barani Dhan-1, DRR Dhan 42, 44, 47,50, Sabour Shree, Birsa VikasDhan 111, Tripura Khara 1
Wheat	A-9-30-1, K 65, 72,78, 7410, 9465, HD 2160, HD1467, N 59, N59, Ajanta, HI 1531, MP (JW) 3173, Ratan, Netravati, HD 2987, HD 3043, KRL-213, PBW 644, WH 1142, Sabour Nirjal, HUW 669, DBW 252, HI 8802, HD3411
Barley	PL419, JB-58, RD 2660, RD-2592
Maize	Mahi Dhawal, JM-216, Pratap Hybrid Makka-1, Pratap Makka-3, Pratap Makka-5, HQPM-1, Bajauramakka, HQPM-5, NAH-1147, PEHM-1
Pearl millet	WCC-75, GHB-538, GHB-538, GHB-719, GHB-719, GHB-757, RHB-154, HHB 223, HHB-216, HHB-226, RHB-177, Bio 70, HHB-234, PBH 306, Balwan, NBH 4903, CZP 9802, Pusa Composite 443, Pusa Hybrid 415
Sorghum	M35-1, CSH-5, CSV-4, CSH-9, Phule Chitra, CSV-17, Pant Chari 5 &7, Phule Panchami, CSH 31R, DSV-2, Phule Vasudha, Parbanimoti, PS-4, SIA-326
Pigeonpea	Palnadu, Co 5, ICPL-87, Maruti, Pragati, Abhay, Jagrati (ICPL 151), Asha, Paras, MAL 13, BRG-1, BRG-2, PRG-158, VL Arhar-1, Rajeev, BDN-711, BDN-716, GT-102, LRG-52, GRG 811, BDN-708, BRG 5, PRG 176, Prakash
Blackgram	Shekhar-2, Azad-3, Pratap Urd-1, (KPU 07-08), BDU-1, Pant Urd-35
Greengram	RMG-268, Pratap, PDM-139, RMG-344, Pusa Vishal, GM-4, CO. (Gg) 8, Yadadri, VBN 4 VGG 10-008, BM-2003-2
Chickpea	JG-315, Phule G-81-1-1 (Vijay), Pant G 186, JG-11, JG -6, JAKI 9218, Akash, RSG 896, JGK-2, JGK-3, LBeG 7, JG 226, JG 14, GNG 1581, BGD 103, Gujarat Junagadh gram 3, AKG 9303-12, PKV Harita (AKG 9303-12), NBeG 3, JG 36, Pant Kabuli gram-2, Pant gram-4, Indira Chana, Nandyal Gram 49, JG 16, Pusa Chickpea 10216
Cowpea	UPC 618, UPC 622, C 519, IT-38956-1, UPC 628, Hidrudaya
Moth bean	CAZRI MOTH-1, RMO-225, RMO-423, CAZRI Moth -3, RMO-257
Horsegram	PHG-9, CRIDA-18R, Indira Kulthi-1, CRIDALATHA, VL Gahat-19 (VLG-19), GHG-5, Bilasa Kulthi (BSP 15-1), Dapoli Kulthi-1
Lentil	Kota Masoor 3 (RKL 605-03), L 4729
Groundnut	TAG-24, Spanish Improved, TMV-1,7, 13; T-28, S-230, GAUG-10, Dh 3-30, 256, 257, Kadiri-3, 5, 7, 9; GG-2, JGN-2, 3, 23, BG-3, MH-4, ICGS-1, 5, 11, 37, 44, 76, ICGV-87160, 86031, 91114, Pragathi, Tirupati-1, 2, 3, 4; Kopergaon-1, Jagtial-88, Mukta, Vemana, TG-26,37A, 51; Smruti, GG-5, Co (gn)- 4, 5, 6, Pratap Mungphali-1, Abhaya, Narayani, Prasuna, AK-265, SG-99, VRI (Gn)-7, R-2001-3, Girnar-2 &3; Greeshma Mallika, Kadiri Haritandhra, GJG-HPS-1, Raj Durga, Girnar-3, R-2001-2, RARS-T-1,2; Divya, Raj Mongfali-1, Dharani, Birsa Groundnut 4, Dheeraj, Kadri lepakshi.
Soybean	Pant Soybean 24, VL SOYA-2, JS-80-21, JS 71-05, Ahilya-3, JS-335, Indira Soya- 9, PS-1092, MACS-450, Parbhani Sona, Ahilya-4, MAU-71, JS-9305, JS-95-60, PS-1225, Pratap Soya-2, RKS-24, Pratap Soya 45 and NRC 136
Sunflower	KBSH-44, JWALAMUKHI (PSCL-5015), MLSFH-47 (AH-11-34), KBSH-41, KBSH-42, DRSF-108, PRORUN-09, Bhanu, KBSH-53, Bhaskar
Safflower	Phule Kusum (JLSF-414), Bhima, Malviya Kusum305 (HUS-305), JSI-7, Parbhani Kusum (PBNS-12), PBNS-40, ISF-764
Niger	Birsa Niger-1, JNC-6, JNC-9, KBN-1, Phule Karala, Birsa Niger-3 (BNS-11)
Castor	GCH-2, GCH-5 (SHB-145, Kranthi (PCS-4), Kiran (PCS-136), 48-1 (Jwala)

Crop	Varieties
Linseed	Jawahar-552, 165, KL-31, NL-97, Shekhar, Sharda, JLS-27 Suyog, Binwa, Indira Alsi-32, Deepika, RLC 92, Himani, JLS-67, JLS-73, Mau Azad Alsi-2
Sesame	GT-10, PB NO.1, TYPE-13, JT-21, TKG-22, JTS-8, Nirmala (OR-Sel-164), Pragati (MT-75), Jawahar Til PKDS-11 (Venket), TKG-306, Amrit, TKG-308, RJ Til -346, G Til-4, RJ Til 351 (RT 351), CUMS-17 (Suprava), RT 372
Rape seed and mustard	Geeta, Pusa Bold, Puas Tarak, Rohini, PT-507, KBS-3, Agrani, Aravali, Karan Tara, Pusa Mustard-21, Divya-33, Pant Rai 20, RH 761, DRMR 150-35
Fruit crops	
Banana	Kaveri Saba, Monthan, Karpuravalley, Poovan, Sugar and Pisang Awak
Citrus	Nagpur mandarin, Nagpur Seedless, PDKV lime, Katol Gold
Pomegranate	Goma Khatta, Ganesh, P-23, P-26, Mridula, Phule Arakta, Bhagwa
Vegetable crops	
Brinjal	PKM-1, KashiSandesh, KashiTaru, KashiHimani
Cassava	H-97, Sree Sahya, Sree Harsha
Chilli	Samrudhi, Kashi Anmol,
Cluster bean	RGC-936, 1017, 1003, 1066, Thar Bhadavi
Dolichos	Arka Jay, Arka Vijay, HA-4
Onion	Agrifound Dark Red, Arka Kalyan, Raseedpura local
Tomato	Arka Meghali, Arka Vikas, Thar Annant
Plantation, medicinal and aromatic crops	
Tea	UPASI-2, UPASI-9, UPASI-20, UPASI-26, BSS-1, BSS-5, TRI-2025, TS 378
Coffee	Selection-7, Selction-9, Selection-11
Coconut	ALR C-1, ALR C-2, Kalpa Dhenu, Kalpa Pratibha, Kalpa Mitra, Kera Keralam, Kalpasamrudhi, Chandra Kalpa, KalpaTaru, Kalpasree, KalpaSankara
Rubber	RRII-430, RRII 105, GL-1, RRII 118 and RRII 203
Black pepper	Panniyur-6, Panniyur-7, Panniyur-8, Panniyur-9 Sigandhini, Pournami
Cardamom	ICRI-5, ICRI-6, ICRI-4
Ginger	CDPK1, Suruchi, Suravi, Himgiri
Turmeric	IISR-Pragathi, CO-2, BSR-1
Coriander	CO-1, CO-2, RCr-20
Fennel	Co-1, GF-1, S-7-9
Ajwain	Ajmer Ajwain-1, Ajmer Ajwain-2, Ajmer Ajwain-93

Table 1.
Drought tolerant crop varieties.

India and Bangladesh. Waterlogging conditions affect plant growth and physiological process and ultimately leading to yield loss. Except for sorghum, most of the field crops are sensitive to waterlogging conditions, and thus it is very crucial to mitigate the effects of waterlogging on crops to sustain food production. Hence, there is tremendous scope for the development of varieties tolerant to excess moisture conditions and which necessities basic research. An extensive basic study at the physiological and molecular level unraveled the traits (aerenchyma and adventitious roots), and mechanisms (formation of a barrier against radial oxygen loss, regulation of ethylene and gibberellic acid, and economizing carbohydrate reserve) associated with submergence tolerance in rice [18].

The most striking progress was the discovery of the SUBMERGENCE 1 (*SUB1*) locus from landrace FR13A and its deployment into popular rice varieties. Consequently, the development of new versions of submergence tolerant and high-yielding popular, and mega rice varieties was witnessed across the world including India [19, 20]. Central Rice Research Institute (ICAR-CRRI), Cuttack and Indian Institute of Rice Research Institute (ICAR-IRRI), Hyderabad in collaboration with international and national institutes developed and released more than 20 submergence-tolerant rice varieties (**Table 2**). Some of them were popularized and subsequently adopted by farmers. The success of the new version of mega varieties was mainly attributed to additional yield and income advantage due to their biotic stress resistant and retention of their original grain quality besides 2-week protection from submergence tolerance [21].

4.3 Heat-tolerant crop varieties

High temperature due to global warming is negatively affecting crop production across the world. High temperatures affect plant growth at various phenological stages, limit biomass production, and mainly reduce the reproductive period to

Sl. No.	Varieties	Organizations/Institute
Rice		
	Sarala, Samba sub-1, Swarna Sub1, CSR43, Reeta, CR Dhan 500, Jalamani, Jayantidhan, CR Dhan505, 801, 802	ICAR-NRRI, Cuttack
1.	Sabour Shree	BAU, Ranchi
2.	MTU 1140, Ksheera	RARS, Maruteru
3.	BahadurSub-1	AAS, RARS, Titabar
4.	DRR Dhan 50	ICAR-IIRR, Hyderabad
5.	Tripura Jala	ICAR RC-NEH, Lembucherra
Wheat		
6.	KRL 19, KRL 210 & KRL 283	CAR-CSSRI, Karnal
Maize		
7.	Jawahar Maize 218	JNKV, Jabalpur
8.	Pusa Jawahar Hybrid Maize-1	IARI, New Delhi
9.	Pragati	GBPUAT, Pantanagar
Pigeonpea		
10.	Maruti, ICPL 84023, Asha	AICRP & ICRISAT
11.	Mal 13	BHU, Varanasi
Soybean		
12.	NRC-37	ICAR-IISS, Indore
13.	JS 97-52	JNKVV, Jabalpur

Table 2.
Crop varieties tolerant submergence and waterlogging.

curtail flower and fruit numbers, thus resulting in severe yield losses. Wheat, one of the global staple food crops, was witnessing severe yield loss in Australia and Pakistan followed by India and China due to heat stress [22] indicating the alarming situation in these countries. This was evident from estimated yield loss [23] and the sliding down of wheat production in India during 2021–2022 due to terminal heat stress. Further, summer crops, such as pulses, coarse cereals, oilseeds, vegetables, and fruits, have also witnessed the heatwaves across northern and western parts of the country.

The adverse effects of heat stress can be mitigated by developing thermotolerant crop varieties through genetic improvement, which demands understanding the mechanisms and traits associated with heat tolerance [24]. Extensive molecular studies and continuous screening efforts in wheat led the identification of QTLs/traits associated with heat tolerance [25, 26] and heat-tolerant genetic resources [27, 28]. Promising wheat varieties such as Parbhani-51 (PBN-51) and Lok-1 designated as heat-tolerant genotypes, are being used in the breeding program beside farmers still cultivating in central India. This shows inherent thermotolerance in old and local varieties and it is supported by information provided in **Table 3**. Recently, number of varieties/hybrids of diverse crops *viz*., wheat (HD 3293, CG 1029, DBW 221, and SHIATS W-13), maize (RCRMH 2), rice (NLR 40024 and Kau Pournami), and mustard (Radhika, Brijraj, and Azad Mahak) tolerant heat stress were released for cultivation in India [29] indicating the scope for heat tolerance breeding. Further, high temperature affects fruit and vegetable crops severely in terms of yield and quality. Thus, efforts are being made by ICAR institutes and other SAUs in developing heat-tolerant varieties in fruits and vegetable crops.

4.4 Cold/frost-tolerant crop varieties

Like heat waves, the occurrence of cold waves is being experienced in the recent past, particularly in northern and western parts of the country, and consideration of cold waves depends on the prevalence of normal temperature in particular regions [14]. Rice and maize, which are staple food crops of India, are highly sensitive to low temperatures as the vegetative and reproductive growth stages are severely affected when temperatures fall below 10°C. Crops that are native to warm environments, exhibit symptoms of injury, reduction of leaf expansion, wilting, chlorosis, and necrosis when subjected to low nonfreezing temperatures. Whereas, severe low temperature for longer periods during pollination or fertilization and seed maturation affect the seed quality and extend harvesting time in agricultural crops besides yield losses. The yield loss due to low temperature depends on the crop stage and severity of stress. For instance, yield loss in sensitive crops such as maize was around 8–10% if low temperature (8–10°C) prevails at the germination stage, where yield penalty was more (60–75%) if the crop was exposed to low temperature (8–10°C) during grain filling stage [30].

Unraveling the crop and even stage-specific cold tolerance mechanisms and traits/genes is very essential for breeding crop varieties tolerant to low temperature [31, 32]. Identification and exploitation of traits/genes associated with cold tolerance helped in developing crop varieties tolerant to low temperature and some of them are listed in **Table 4**. The increasing trend of cold-tolerant varieties in rice, wheat, mustard, and other forage and vegetable crops indicates the increasing incidence of the negative effect of low temperature on crop yield and quality irrespective of growing seasons and crop habitat. Thus, the identification of genetic stocks or germplasm, genes or QTLs, and other molecules including nucleotide sequences and proteins associated with cold tolerance and their utilization in the development of cold-tolerant varieties

Sl. No.	Varieties	Year	Organizations/Institute
Rice			
1.	DRR Dhan 47 & 52	2019	ICAR-IIRR, Hyderabad
Wheat			
2.	Lok-1	1981	Lokbharati Foundation
3.	Ajantha	1983	CoA, Badnapur
4.	HD 2402	1988	IARI, New Delhi
5.	K9465 and HD 2160	1998	CSAUA&T, Kanpur
6.	HW 2045	2002	IARI, New Delhi
7.	Pusa gold	2003	
8.	RAJ 3765	2006	RARI, Durgapura
9.	DWR17	2007	ICAR-IIWBR, Karnal
10.	K 0307	2007	CSAUA&T, Kanpur
11.	WH 1124	2014	ICAR-IIWBR, Karnal
12.	DBW 107	2015	
13.	DBW 173	2018	
14.	HI 1634 (Pusa Ahilya)	2021	ICAR-IARI, Regional Station, Indore
Maize			
15.	Suwan	-	BAU, Sabour
16.	PMH-7	-	PAU,Ludhaiana
17.	RCRMH2	2016	UAS, Raichur
Pearl millet			
18.	GHB-558,538, 732	-	JAU, Jamnagar
Pigeonpea			
19.	Bahar	1986	RAU, Pusa
20.	UPAS-120	1976	GBPUA&T, Pantanagar
Chickpea			
21.	Indira Chana	-	IGKV, Raipur
22.	JG-315	1984	JNKV, Jabalpur
23.	Pant G 186	1996	GBPUAT, Pantanagar
24.	JG-11	1999	JNKV, Jabalpur
25.	JG -6, JAKI 9218	2006	
26.	JG-14	2009	
Moth bean			
27.	RMO-40	1994	ICAR-CAZRI, Jodhpur
28.	RMO-225, CAZRI Moth-1	1999	
Lentil			
29.	Kota Masoor 3	2020	AU, Kota
Groundnut			

DOI: http://dx.doi.org/10.5772/intechopen.1001916

Sl. No.	Varieties	Year	Organizations/Institute
30.	Kadiri-3	1978	APAU, Kadiri
31.	Phulepragati (JL-24)	1978	MPKV, Jalgaon
32.	ICGS-44 , 76	1988	ICRISAT, Hyderabad
33.	ICGV-86031	1991	
34.	TG-22	1992	BARC, Mumbai
35.	Ambar (CSMG-84-1)	1992	CSAUA&T, Mainpuri
36.	Vemana (K-134)	1995	APAU, Kadiri
37.	Kadiri-6 (K-1240)	2005	ANGRAU, Kadiri
Soybean			
38.	JS-335	1994	JNKVV, Jabalpur
Sunflower			
39.	DRSF-113	2007	ICAR-IIOR, Hyderabad
40.	PSFH-118	2002	PAU, Ludhiana.
41.	Jawahar-552 (R-552)	1980	IGKV, Jabalpur
42.	Deepika (RLC-78)	2006	
43.	Indira Alsi-32 (RLC-81)	2005	KGKV, Raipur
Linseeds			
44.	RLC 92 (IC 555926)	2008	IGKV, Jabalpur
Sesame			
45.	CUMS-17 (Suprava)	2018	IAS, University of Calcutta
Rape seed and mustard			
46.	NRCDR-02	2007	ICAR-DRMR, Bharatpur
47.	Pusa Mustard 22 & Vijay	2008	IARI, New Delhi
48.	Pusa Tarak	2009	
49.	Pusa mustard 25 (NPJ 112)	2010	
50.	NRCDR 601 (DMR 601)	2010	DRMR, Bharatpur
51.	RGN 229 and RGN 236	2011	RAU,ARS, Sriganganagar
52.	Pusa Mustard 27(EJ 17)	2011	IARI, Regional Station, Karnal
53.	Pusa Mustard 26 & 28	2012	IARI, Karnal
54.	Pant Rai-19 (PR-2006-1)	2012	GBPAUT, Pantanagar
55.	Divya-33	2013	M.R. Seeds Pvt. Ltd.
56.	Pusa mustard 29 & 30	2013	IARI, New Delhi
57.	RH 0406	2013	CCSHAU, Hissar,
58.	Pro 5222 Bayer Mustard	2019	Bayer BioScience Pvt. Ltd.
59.	Vasundhara		CCS HAU, HISSAR
Vegetable crops			
Bottle gourd			
60.	Thar Samridhi		CIAH, Bikaner

Sl. No.	Varieties	Year	Organizations/Institute
61.	Pusa Santushti	2005	IARI, New Delhi
Brinjal			
62.	Kashi Sandesh,	2004	IIVR, Varanasi
63.	Kashi Taru	2005	
64.	Thar Rachit		CIAH, Bikaner
Chilli			
65.	Kashi Abha	2019	IIVR, Varanasi
Cucumber			
66.	PusaBarkha	2012	IARI, New Delhi
Okra			
67.	KashiPragati	2004	IIVR, Varanasi
68.	KashiKranti	2012	
Potato			
69.	Kufri Surya Surya	-	CPRI, Shimla
70.	Kufri Lauvkar	-	
Radish			
71.	Pusa Chetaki	1988	IARI, New Delhi
72.	Kashi Mooli-40	2019	IIVR, Varanasi
Ridge gourd			
73.	Thar Karni	-	CIAH, Bikaner
74.	Pusa Sneha	2004	IARI, New Delhi
75.	Thar Tapish	2018	CIAH, Bikaner
Tomato			
76.	Pusa Sadabahar	2004	IARI, New Delhi
77.	Arka Meghali	2006	IIHR, Bengaluru
78.	Varkha Bahar-1 & 2	2009	PAU, Punjab
79.	Arka Vikas	-	IIHR, Bengaluru
Water melon			
80.	Thar Manak	2016	CIAH, Bikaner

Table 3.
Crop varieties tolerant to high temperature.

is very crucial. Similar to heat-tolerant varieties, most of the reported genotypes tolerant to low temperature are old and local varieties **Table 4**.

4.5 Salt-tolerant crop varieties

Salt stress particularly, salinity is becoming a major threat to crop production in arid and semiarid regions of the country where deficit moisture is a common scenario and also in the irrigated agricultural lands with inadequate drainage. The

Sl. No.	Varieties	Year	Organizations/Institute
Rice			
1.	Kalinga 1	1973	ICAR-NRRI, Cuttack
2.	Tellahamsa	1975	ANGRAU, Hyderabad
3.	Pant Dhan 11	1993	GBPUAT, Pantanagar
4.	Varun Dhan	2008	CSKHPKV, Palampur
5.	Gizza-14, K-39, 343 & 448	-	SKUA&T, Jammu
6.	NE Megha Rice 1 & 2	-	ICAR Barapani
Wheat			
7.	Buland		PAU, Ludhiana
8.	Mansarovar	1999	SKUAST, Srinagar
9.	Shalimar wheat-1	2005	
10.	RSP 561	2015	
Pearl millet			
11.	GHB-538	2005	JAU, Jamnagar
Pigeonpea			
12.	Bahar	1986	RAU, Pusa
Chickpea			
13.	PDG 4	2003	GBPUAT, Pantanagar
Rape seed and mustard			
14.	RH-9801 (SWARNA)	2003	CCS HAU, HISSAR
15.	RGN-48, 73	2007	RAU, ARS, Sriganganagar.
Buffalo grass			
16.	Bundel Anjan-1	1989	IGFRI, Jhansi
Dhaman			
17.	Bundel Dinanath-2	1989	IGFRI, Jhansi
Rye grass			
18.	Pb. Ryegrass No.1	1991	PAU, Ludhiana
Setaria grass			
19.	PSS-1, Nandi	1983	CSK HPKVV, Palampur
20.	Setaria-92	2005	
21.	S-18	2013	ICAR-IGFRI, Jhansi
22.	Him Palam Setaria - 2 (S -25)	2020	CSK HPKVV, Palampur
Bottle gourd			
23.	Pusa Santushti	2005	IARI, New Delhi
Carrot			
24.	Ooty-1	1997	TNAU, Coimbatore
25.	PusaYamdagni	1986	IARI, New Delhi
Knol-khol			

Sl. No.	Varieties	Year	Organizations/Institute
26.	Pusa Sheetal	1987	IARI, New Delhi
27.	Pusa Nayanjyoti & Pusa Virat	2009	
Tomato			
28.	Pusa Sadabahar	2004	IARI, New Delhi

Table 4.
Crop varieties tolerant to low temperature (cold/frost).

Sl. No.	Varieties	Organizations/Institute
	Rice	
1.	Vikas DRR Dhan 9, 33, 39	ICAR-IIRR, Hyderabad
2.	CSR10, CSR13, CSR23, CSR27, CSR30, CSR36, CSR43, CSR46, CSR49, CSR52, CSR56, CSR60 and CSR76	ICAR-CSSRI, Karnal
3.	Luna Suvarna, CR Dhan 402, 403, 405, 406	ICAR-NRRI, Cuttack
4.	Panvel 3	BSKV, Dapoli
5.	CARI Dhan 5	CARI , Port Blair
Wheat		
6.	KRL 1-4 , KRL 19, KRL 210, KRL 283, KRL-213	ICAR-CSSRI, Karnal
Pigeon pea		
7.	UPAS 120	GBPUAT, Pantanagar
8.	Jagriti (ICPL 151)	ICRISAT, Hyderabad
9.	C 11	ANGRAU, Hyderabad
Chickpea		
10.	Karnal chana-1	ICAR-CSSRI, Karnal
Cowpea		
11.	Hissar Cowpea-46	CCS, Hissar
12.	Bidhan Rice bean-3	BCKV, West Bengal
Rice bean		
13.	Vikrant (VH-82-1)	CCS, Hissar
Lentil		
14.	PSL-9 and PDL-1	ICAR-CSSRI Karnal
Rape seed and Mustard		
15.	CS 52, CS 54, CS 56, CS 58, CS 60	ICAR-CSSRI Karnal
16.	Pusa Vijay, Pusa mustard 25, 22, 29 & 30	IARI, New Delhi
17.	NRCDR 601 & NRCDR-02, Giriraj (DRMRIJ 31)	DRMR, Bharatpur
18.	RGN 229, RGN-145 & RGN 236	RAU, Sriganganagar
19.	RH0119, RH 0406	CCS HAU, HISAR
20.	Narendra Rai	NDAUT Faizabad.

Table 5.
Salt stress tolerant crop varieties.

intrusion of seawater in the coastal area is another foremost root cause for the increase in salinity. Sodicity in clay soils is a consequence of salinity, where leftover over-soluble salts are leached into the subsoil and sodium is left bound to the negative charges of the clay [33]. Sodic soils are more common in the pockets of the arid and semiarid regions of western and central India and peninsular tracts in southern India beside patches of Indo-Gangetic plains. On the other hand, acid soils are problematic soils formed by human activities, such as construction and mining and highly acidic soils naturally prevailed in the Himalayan ecosystem, red, and lateritic regions of India [34]. Increased salinization and sodicity and acidity of arable lands are expected to have devastating effects on agricultural production as these soil conditions cause hyper-ionic and hyperosmotic stresses in plants and affect normal growth and developmental process [35]. Whereas, acid soils affect plant growth indirectly through the dissolution of harmful elements and causing nutrient deficiency besides directly affecting the plant metabolic process due to an imbalance in pH [36]. Though data on crop yield loss associated with the type of salt-affected soils may not be available it is evident from the available literature that there was significant yield loss of up to 34 and 69% in major staple crops such as maize due to salinity and acidity, respectively [30, 37].

To cope with the effects of salt stress, the development stress-tolerant crop varieties is an eco-friendly and economically viable option, particularly in the regions where the above-mentioned problematic soils spread across a large area and are difficult to reclaim with limited available resources. Breeding varieties tolerant to specific problematic soils pose a great challenge for plant breeders, which necessitates unraveling the mechanisms and traits/genes associated with salinity, sodicity, and acidity tolerance [37, 38]. A number of mapping studies have been attempted to identify candidate genes or QTLs for salinity tolerance in different crops including rice [39–43], wheat [44], and chickpeas [45]. However, limited progress has been made in the development of salt-tolerant cultivars. As a premier institute of ICAR, CSSRI, Karnal working in the area of crop improvement for problematic soils has developed and released varieties in rice (13), wheat (5), mustard (5), and chickpea (1) for saline and acid soils (**Table 5**). For instance, several rice breeding lines were developed through marker-assisted breeding by transferring *Saltol* and *qSSISFHS8.1* QTLs in the genetic background of Indian mega rice varieties *viz*., Pusa44, Sarjoo52, and PR114 [46].

In fruit crops, there is scope for the identification of rootstocks tolerant to salinity as these crops also witness the effects of salt stress on yield and quality under the scenario of accelerated climate change. Several rootstocks in different fruit crops were identified across the world [47, 48], and some of them were collected and maintained by NARS institutes (**Table 6**).

5. Reorientation of breeding strategies to develop climate-smart varieties: A way forward

Most of the ASTCVs enlisted and discussed in the previous section were developed through conventional breeding approaches by exploiting heritable, adaptive phenological traits such as early flowering/maturity, root traits, and other morphological traits. Except a few salinity-tolerant crop varieties and submergence-tolerant rice varieties, several of ASTCVs are not being cultivated by farmers on a large scale. This is mainly due to a lack of awareness and availability of quality seeds of such

Sl. No.	Crop	Genotypes/varieties/lines/genetic stocks	Stress
1.	Apple	M4,7,26,104,106, 111 *Malus baccata*, *M. toringoides*, *M. sieversii*	Drought
		G 11, 30, 41, 935, B 9, P 2, M7, 26, 106,	Cold
2.	Apricot	*Prunus armeniaca* 'Suka' wild type	Cold
		P. armeniaca 'Sahara' wild type	Heat
		Line FA-3-6, Pollizo prune	Waterlogging
3.	Avocado	West Indian race	Salinity
4.	Blue berry	*V. corymbosum* x V. *angustifolium*	Cold
		V. darrowi, *V. amoenum*, *V. arboretum*, *V. elliottii*, *V. myrsinites*	Heat
5.	Cherry	*P. avium*	Drought
		Prunus cornuta, *Prunus cerasoides*	Cold
6.	Citrus	Rangpur lime, Marmlade orange, Rough Lemon, Cleopatra mandarin, Sour Orange, Karna Khatta, Nasnaran, attani 2, Gou Tou Cheng	Salinity & Drought
7.	Guava	Crioula'	Salinity
8.	Grapes	*V.champini*, *V. vinifera*, Dogridge, salt creek, *V. berlandieri* × *V. rupestris*, 1613, 1103 Paulsen	Salinity
9.	Kiwi fruit	*A. valvata* (KR5 & KR3)	Waterlogging
10.	Mango	Nekkare, Terpentine, Olour, Bappakai, Gomera-1, *M. zeylanica*, ML-2, GPL-1	Salinity
11.	Peach	GF677, 577, Myrobalan, Bright Hybrid	Drought & Salinity
12.	Pear	*Pyrusbetulifolia*	Salinity
		Oregon 211 and 249, Oregon 260, 261, 264, Quince A, Provence Quince,	Drought
13.	Pistachionut	*P. atlantica*, UCB-1	Salinity
14.	Plum	Mariana 4001, Peach- Almond Hybrid, Myrobalan 27, GF 667 and GF 577	Drought
15.	Pomegranate	Tab-o-Larz	Salinity

Table 6.
Rootstocks of fruit crops tolerant to different abiotic stresses.

varieties. Further, the poor and variable genetic potential of varieties was another major limitation, resulting in poor quality and performance in comparison to other high-yielding varieties (HYVs) under normal conditions besides being susceptible to pests and diseases. Therefore, it is very essential to develop climate-smart varieties with higher genetic gain in terms of yield and quality and should perform irrespective of any stress/es or nonstress condition. However, breeding, such varieties, is a very big challenge since abiotic stress tolerance is a complex trait and is controlled by several genes/quantitative trait loci (QTLs). With novel discoveries and the faster development of many new techniques and tools, great progress has been made in the last decade in delivering scientific leads for understanding the abiotic stress tolerance mechanisms in plants. However, such advanced tools and techniques failed to translate novel findings into deliverable and significant outputs in the form climate-smart varieties. It is very crucial to adopt new tools in the right way and reorient the breeding strategies to gear up a genetic gain of crop plants to breed climate smart with multiple stress-tolerant varieties.

5.1 Genomic interventions

The current understanding of abiotic tolerance in crop plants reveals the general role of some regulatory factors of gene expression involved in stress tolerance mechanisms. Further, in the last two decades identified the common QTLs/genes/traits involved in multiple stress tolerance through extensive and interdisciplinary molecular studies. Identification and characterization of these factors by exploiting the new and advanced molecular tools will not only deliver the scientific leads to the understanding complex mechanism of abiotic stress tolerance. Subsequently, it helps to develop stable and climate-smart crop varieties through integrated molecular-conventional breeding approaches. For instance, several QTLs associated with drought tolerance have been identified and exploited in breeding program in chickpea [49], groundnut [50], maize [30], etc. Further, forward genetics tools including fine mapping, map-based cloning, TILLING, and eco-TILLING are being exploited to unravel the candidate genes controlling the stress tolerance and their functions at the genome level. In addition, together with microarray technology affordable sequencing tools enables decoding the sequences of coding regions of the genome regulating the stress tolerance even at tissues specific level. The new and cutting-edge molecular techniques, such as RNAi, CRISPR/CAS, or DNA-free genome editing (DFFE), are more precise and reduce the risk of off-targets, which is commonly encountered in the transgenic approach.

The metadata information from advanced 'genomic' tools provides insight for the identification of 'hot spots' or 'genomic regions' controlling the stress tolerance mechanisms. In addition, unraveling the expression networks of stress-responsive, defense signaling, and expression genes encoding for proteins and enzymes involved in cellular-level defense mechanisms, which ultimately results in tolerance at the phenotypic level. The exploration of such insights in the integrated and physiological-based breeding program accelerates genetic gains in abiotic stress resistance breeding. However, the identification of candidate genes/QTLs or network or regulatory genes governing the abiotic tolerance remains as a great challenge due to the large number of genes influencing the targeted traits. Hence, an extensive and continuous investigation into the basis of tolerance in model crops or arid and semiarid crops, xerophytes, and halophytes will undoubtedly provide a clearer insight into the genes/traits associated with different abiotic stress tolerance mechanisms.

5.2 Phenomic interventions

Dissecting and quantifying the effects of genotypes (G), phenotype (P), and environment (E) components and subsequent utilization of comprehensive information is key for unlocking the complex genetics mechanism of trait/s. This approach holds good for biotic and abiotic stress tolerance mechanisms in crops. Even after dissecting the genotype of plants using modern and advanced tools, which are discussed in the above section, understanding a clear picture of phenotype is very crucial and it stands foremost among all the three components. Hence, since last decade crop phenotyping communities across the world being witnessed the use of reliable, automatic, multifunctional, and high-throughput phenotyping technologies for phenotyping and subsequent rapid advancement of genetic gain in breeding programs. In this context, the plant phenotyping experts not only designing the multi-domain, multi-level, and multi-scale crop phenotyping models [51] to generate a big database but also continuously making efforts to standardize precise protocols for the identification of phenotypic traits [52] and develop bioinformatics tools for mining information from the exhaustive omics data [53].

5.3 Integrating omics tools with conventional and molecular breeding

Understanding the genetic, molecular, and physiological complex mechanisms of abiotic stress tolerance, which are interconnected through biochemical or gene networks by integrating different omics (genomics, proteomics, and metabolomics) with bioinformatics tools is very crucial for developing climate-smart varieties. In this context, phenomics and modeling communities together with plant scientists can exchange interdisciplinary knowledge through a common platform and multi-omics data, through intelligent data-mining analysis [54], which offers a powerful tool to unravel the physio-morphological responses upon multiple stresses [55] and tolerance mechanisms thereof much needed for the development of climate-smart varieties. Thus, simplifying the complexity of biological processes by dissecting genetics or molecular action (including pleiotropy, epigenetics if any) of the QTLs/genes or biomolecules associated with abiotic stress tolerance is a basic necessity for climate-smart breeding. This provides a holistic understanding of heritability, expression, and penetrance of traits or genes controlling stress tolerance. Further, phenotyping the candidate trait/gene or its surrogated traits through high-throughput phenotyping tools and quantitative genetic models enables to confirm the expression of traits/genes by measuring trait values at the phenotypic level in a large number of diverse germplasm or mapping population. Finally, the genetic enhancement of various crops can be carried out through the transfer of stable and candidate traits/QTLs/genes using different approaches such as marker-assisted selection (MAS), marker-assisted recurrent selection (MARS), genome-wide selection (GWS), and genetic engineering/transgenics.

In view of the remarkable progress in 'omics' areas in recent years, the integration of biotechnological approaches such as molecular breeding with conventional breeding along with different omics tools (particularly genomics and phenomics) should be the major emphasis for accelerating the development climate-smart varieties. Overall, linking biophysical and genetic models to integrate physiology, molecular biology, and plant breeding helps in dissecting or unraveling the complex traits associated with abiotic stress tolerance which accelerates the breeding program to develop climate-smart varieties with great resilience to abiotic stresses [56]. In recent years, remarkable progress has been made by plant breeders of NARS in this direction and succeeded in the development of drought-tolerant varieties in chickpea [49] and wheat [57] through marker-assisted breeding.

6. Conclusion

Mitigating the impact of abiotic stresses, which are accelerated by climate change is very crucial to safeguard national food and nutritional security, particularly in the arid, semiarid, and coastal regions. The cultivation of abiotic stress-tolerant crop varieties (ASTCVs) is one of the best strategies to cope up climate change and consequent abiotic stresses. Indian NARS bred a number of climate-resilient varieties tolerant to different biotic and abiotic stress-tolerant varieties. Documentation of ASTCVs with success stories of adoption of such varieties is very important to popularize and create awareness on climate-resilient varieties. Further, developing strategic plans to enhance the availability quality seeds of is a key factor for reaping the benefits of trait values of climate-resilient varieties. Most of the ASTCVs are developed accidentally but not exclusively by abiotic resistance breeding program, and thus only a few of them, which are stable and perform in both stress and normal conditions. Hence it is very crucial to shift from climate-resilient varieties (ASTCVs) to climate-smart varieties, tolerant multiple biotic, and abiotic stresses. Accordingly, national and international organizations reoriented their breeding strategies to develop climate-smart varieties through an integrated 'omic' approach. Along the way, continuous efforts are being made to understand mechanisms/traits associated with different abiotic stress tolerance by exploiting the advanced tools and techniques in molecular science. The smart breeding comprising modern pre-breeding, gene/QTLs mapping using MAGIC population, genomic-based selection and mapping, genetic engineering approach with advanced genome editing, and high throughput genotyping and phenotyping (HTGP) tools along with speed breeding technique accelerates the abiotic resistance breeding program. This integrated approach helps in the development of climate-smart varieties within a short period of time to safeguard food and nutritional security in the abiotic stress-prone regions of the country through enhanced and sustainable food production systems under changing climate scenarios.

Acknowledgements

Authors are thankful to ICAR institutes for providing the information.

Funding

No funding is available for this publication.

Conflict of interest

Authors declare there is no conflict of interest.

Notes

Authors are not responsible for any damage caused by the content/information of this chapter.

Author details

Boraiah K.M.[1*], Basavaraj P.S.[1], Vijaysinha D. Kakade[1], Harisha C.B.[1], Pratapsingh Khapte[1], Halagundegowda G.R.[2], Krishnamurthy D.[3], Neeraj Kulshreshtha[4], Vijayakumar H.P.[5], Bhojaraj Naik[6], Jagadish Rane Sammi Reddy K.[1] and Himanshu Pathak[7]

1 ICAR-National Institute of Abiotic Stress Management, Baramati, Pune, Maharashtra, India

2 Central Silk Board, Bengaluru, India

3 Tieerra Seed Science Pvt. Ltd. Hyderabad, India

4 ICAR- Central Soil Salinity Research Institute, Karnal, India

5 ICAR-Indian Institute of Agriculture Research, New Delhi, India

6 ICAR-Indian Institute of Seed Science, Regional Station, Bengaluru, India

7 Department of Agriculture Research and Education (DARE), Indian Council of Agriculture Research (ICAR), New Delhi, India

*Address all correspondence to: bors_km@yahoo.co.in

References

[1] Pathak H, Mishra JP, Mohapatra T. Indian Agriculture after Independence. New Delhi: Indian Council of Agricultural Research; 2022. p. 426

[2] Pathak H, Aggarwal PK, Singh SD. Climate Change Impact, Adaptation and Mitigation in Agriculture: Methodology for Assessment and Applications. New Delhi, India: Indian Agricultural Research Institute; 2012

[3] Venkateswarlu B. Climate change scenario in India and its impact on agroecosystems. In: Chary R, Srinivasa Rao G, Srinivas K, Maruthi Sankar GR, Nagarjuna Kumar R, Venkateswarlu B, editors. Adaptation and Mitigation Strategies for Climate Resilient Agriculture. ICAR, Hyderabad, India: Central Research Institute for Dryland Agriculture; 2013. pp. 1-16

[4] Mukherjee S, Aadhar S, Stone D, Mishra V. Increase in extreme precipitation events under anthropogenic warming in India. Weather and Climate Extremes. 2017;**30**:1-9

[5] Ahluwalia VK, Malhotra S. Environmental Science. New Delhi: Anne Books India; 2006

[6] Kumar NS, Aggarwal PK, Rani SD, Saxena R, Chauhan N, Jain S. Vulnerability of wheat production to climate change in India. Climate Research. 2014;**59**(3):173-187. DOI: 10.3354/cr01212

[7] Rane J, Singh AK, Kumar M, Boraiah KM, Meena KK, Pradhan A, et al. The adaptation and tolerance of major cereals and legumes to important abiotic stresses. International Journal of Molecular Sciences. 2021;**22**(23):12970. DOI: 10.3390/IJMS222312970

[8] Sharm S, Manjeet. Heat stress effects in fruit crops: A review. Agricultural Reviews. 2020;**1**(1):73-78

[9] Seo HJ, Sawant SS, Song J. Fruit cracking in pears: Its cause and management—A review. Agronomy. 2022;**12**:2437. DOI: 10.3390/agronomy12102437

[10] Ismail AM, Singh US, Singh S, Dar MH, Mackill DJ. The contribution of submergence-tolerant (Sub1) rice varieties to food security in flood-prone rainfed lowland areas in Asia. Field Crops Research. 2013;**152**:83-93. DOI: 10.1016/j.fcr.2013.01.007

[11] Chattopadhyay K, Sukanta G, Ismail M, Sumanta M, Mukherjee Arup K, Marandi BC, et al. Impact of climate resilient varieties on Rice productivity and ensuring food security in coastal area of eastern India. In: NRRR Research Bulletin No. 10. Cuttack, Odisha, India: National Rice Research Institute; 2016. p. 68

[12] Dar MH, Waza SA, Shukla S, Zaidi NW, Nayak S, Hossain M, et al. Drought tolerant Rice for ensuring food security in eastern India. Sustainability. 2020;**12**:2214. DOI: 10.3390/SU12062214

[13] Radeny M, Rao EJZ, Ogada MJ. Impacts of climate-smart crop varieties and livestock breeds on the food security of smallholder farmers in Kenya. Food Security. 2022;**14**:1511-1535. DOI: 10.1007/s12571-022-01307-7

[14] Boraiah KM, Basavaraj PS, Harisha CB, Kochewad SA, Khapte PS, Bhendarkar MP, et al. Abiotic stress tolerant crop varieties, livestock breeds and fish species. In: Technical Bulletin No. 32. Baramati, Pune, Maharashtra,

India: ICAR-National Institute of Abiotic Stress Management; 2021. p. 83

[15] Kumar A, Dixit S, Ram T, Yadaw RB, Mishra KK, Mandal NP. Breeding high-yielding drought-tolerant rice: Genetic variations and conventional and molecular approaches. Journal of Experimental Botany. 2014;**65**(21):6265-6278

[16] Abady S, Shimelis H, Janila P, Y aduru S, Shayanowako AIT, Deshmukh D. Assessment of the genetic diversity and population structure of groundnut germplasm collections using phenotypic traits and SNP markers: Implications for drought tolerance breeding. PLoS One. 2021;**16**(11):e0259883

[17] Yamano T, Dar MH, Panda A, Gupta I, Malabayabas ML, Kelly E. Impact and Adoption of Risk-Reducing Drought-Tolerant Rice in India. 3ie Impact Evaluation Report 72. New Delhi: International Initiative for Impact Evaluation (3ie); 2018

[18] Xu K, Xu X, Fukao T, Canalas P, Maghirang-Rodriguez R, Heuer S, et al. Sub1A is an ethylene responsive-factor-like gene that confers submergence tolerance to rice. Nature. 2006;**442**:705-708

[19] Septiningsih EM, Pamplona AM, Sanchez DL, Neeraja CN, Vergara GV, Heuer S, et al. Development of submergence-tolerant rice cultivars: The Sub1 locus and beyond. Annals of Botany. 2009;**103**(2):151-160. DOI: 10.1093/aob/mcn206 Epub 2008 Oct 30

[20] Pathak H, Parameswaran HN, Subudhi SR, Prabhukarthikeyan PSK, Anandan A, Yadav MK, et al. Rice varieties of NRRI: Yield, quality, special traits and tolerance to Biotic & Abiotic Stresses. In: NRRI Research Bulletin No. 20. Cuttack, Odisha, India: ICAR-National Rice Research Institute; 2019. p. 68

[21] Anonymous, Annual Report, 2019-20. Indian council of Agriculture Research (ICAR). New Delhi, India

[22] Zulkiffal M, Ahsan A, Ahmed J, Musa M, Kanwal A, Saleem M, et al. Heat and drought stresses in wheat (*Triticum aestivum* L.): Substantial yield losses, practical achievements, improvement approaches, and adaptive mechanisms. In: Plant Stress Physiology. London, UK: IntechOpen; 2020. DOI: 10.5772/intechopen.92378

[23] Dubey R, Pathak H, Chakrabarti B, Singh S, Gupta DK, Harit RC. Impact of terminal heat stress on wheat yield in India and options for adaptation. Agricultural Systems. 2020;**181**:102826. DOI: 10.1016/j.agsy.2020.102826

[24] Ni Z, Li H, Zhao Y, Peng H, Hu Z, Xin M, et al. Genetic improvement of heat tolerance in wheat: Recent progress in understanding the underlying molecular mechanisms. Crop Journal. 2017;**6**(1):2-4

[25] Paliwal R, Röder MS, Kumar U, Srivastava JP, Joshi AK. QTL mapping of terminal heat tolerance in hexaploid wheat (*T. aestivum* L.). Theoretical and Applied Genetics. 2012;**125**(3):561-575

[26] Kushwah A, Bhatia D, Singh I, Thudi M, Singh G, Bindra S. Identification of stable heat tolerance QTLs using inter-specific recombinant inbred line population derived from GPF 2 and ILWC 292. PLoS One. 2021;**16**(8):e0254957

[27] Mishra SC, Singh SK, Patil R. Breeding for heat tolerance in wheat. In: Shukla RS, Mishra PC, Chatrath R, Gupta RK, Tomar SS, Sharma I, editors.

Recent Trends on Production Strategies of Wheat in India. Karnal, India: JNKVV, Jabalpur & ICAR-IIWBR; 2014. pp. 15-29

[28] Chaudhary S, Devi P, Bhardwaj A, Jha UC, Sharma KD, Prasad PVV, et al. Identification and characterization of contrasting genotypes/cultivars for developing heat tolerance in agricultural crops: Current status and prospects. Frontiers in Plant Science. 2020;**11**:587264. DOI: 10.3389/fpls.2020.587264

[29] Anonymous, Annual Report, 2021-22. Indian Council of Agriculture Research (ICAR). New Delhi, India: DARE-ICAR; 2022

[30] Boraiah KM, Basavaraj PS, Halli HM, Muhkri G, Yathish KR, Ranjan N, et al. Maize: Impacts and management of abiotic stresses. In: Pathk H et al., editors. Abiotic Stresses in Agriculture: Impacts and Management. Baramati, Pune, Maharashtra, India: ICAR-National Institute of Abiotic Stress Management (ICAR-NIASM); 2022. pp. 116-150

[31] Sanghera GS, Wani SH, Hussain W, Singh NB. Engineering cold stress tolerance in crop plants. Current Genomics. 2011;**12**(1):30-43. DOI: 10.2174/138920211794520178

[32] Adhikari L, Baral R, Paudel D, Min D, Makaju SO, Poudel HP, et al. Cold stress in plants: Strategies to improve cold tolerance in forage species. Plant Stress. 2022;**4**:100081. DOI: 10.1016/J.STRESS.2022.100081

[33] Wang W, Vinocur B, Altman A. Plant responses to drought, salinity and extreme temperatures: Towards genetic engineering for stress tolerance. Planta. 2003;**218**:1-14

[34] Maji AK, Obi Reddy GP, Sarkar D. Acid Soils of India - Their Extent and Spatial Variability, NBSS Publication No. 145, NBSS&LUP, Nagpur. 2012. p. 138

[35] Munns R, Tester M. Mechanisms of salinity tolerance. Annual Reviews of Plant Biology. 2008;**59**:651-681

[36] Matsumoto S, Shimada H, Sasaoka T, Miyajima I, Kusuma GJ, Gautama RS. Effects of acid soils on plant growth and successful revegetation in the case of mine site. In: Oshunsanya S, editor. Soil pH for Nutrient Availability and Crop Performance. London, UK: IntechOpen; 2017. DOI:10.5772/intechopen.70928

[37] Ngoune Tandzi L, Mutengwa CS, Ngonkeu ELM, Gracen V. Breeding maize for tolerance to acidic soils: A review. Agronomy. 2018;**8**(6):84. DOI: 10.3390/agronomy8060084

[38] Hanin M, Ebel C, Ngom M, Laplaze L, Masmoudi K. New insights on plant salt tolerance mechanisms and their potential use for breeding. Frontiers in Plant Science. 2016;7:1787. DOI: 10.3389/fpls.2016.01787

[39] De Leon TB, Linscombe S, Subudhi PK. Identification and validation of QTLs for seedling salinity tolerance in introgression lines of a salt tolerant rice landrace 'Pokkali'. PLoS One. 2017;**12**(4):e0175361. DOI: 10.1371/journal.pone.0175361

[40] Puram VRR, Ontoy J, Subudhi PK. Identification of QTLs for salt tolerance traits and pre-breeding lines with enhanced salt tolerance in an introgression line population of rice. Plant Molecular Biology Reporter. 2018;**36**:695-709. DOI: 10.1007/s11105-018-1110-2

[41] Mazumder A, Rohilla M, Bisht DS. Identification and mapping of quantitative trait loci (QTL) and

epistatic QTL for salinity tolerance at seedling stage in traditional aromatic short grain rice landrace Kolajoha (*Oryza sativa* L.) of Assam, India. Euphytica. 2020;**216**:75. DOI: 10.1007/s10681-020-02602-0

[42] Nakhla WR, Sun W, Fan K, Yang K, Zhang C, Yu S. Identification of QTLs for salt tolerance at the germination and seedling stages in Rice. Plants. 2021;**10**(3):428. DOI: 10.3390/plants10030428

[43] Rathor S, Krishnamurthy SL, Lokeshkumar BM, Warraich AS, Yadav S, Sharma PC, et al. Dissection of genomic regions for ion homeostasis under sodic salt stress in MAGIC rice population. Environmental Sciences Proceedings. 2022;**16**:39. DOI: 10.3390/environsciproc2022016039

[44] Asif MA, Garcia M, Tilbrook J, Brien C, Dowling K, Berger B, et al. Identification of salt tolerance QTL in a wheat RIL mapping population using destructive and non-destructive phenotyping. Functional Plant Biology: FPB. 2021;**48**(2):131-140. DOI: 10.1071/FP20167

[45] Soren KR, Madugula P, Kumar N, Barmukh R, Sengar MS, Bharadwaj C, et al. Genetic dissection and identification of candidate genes for salinity tolerance using axiom®CicerSNP Array in chickpea. International Journal of Molecular Sciences. 2020;**21**(14):5058. DOI: 10.3390/ijms21145058

[46] Krishnamurthy SL, Lokeshkumar BM, Rathor S, Warraich AS, Yadav S, Gautam RK, et al. Development of salt-tolerant Rice varieties to enhancing productivity in salt-affected environments. Environmental Sciences Proceedings. 2022;**13**:30. DOI: 10.3390/environsciproc2022016030

[47] Duran VH, Raya AM, Aguilar J. Salt tolerance of mango rootstocks (*Magnifera indica* L. cv. Osteen). Spanish journal of. Agricultural Sciences. 2003;**1**:68-78

[48] Sá FVDS, Nobre RG, Silva LD, Moreira RC, Paiva EP, FAD OA. Tolerance of guava rootstocks under salt stress. Revista Brasileira de Engenharia Agrícola e Ambiental. 2016;**20**:1072-1077

[49] Bharadwaj C, Tripathi S, Soren KR, Thudi M, Singh RK, Sheoran S, et al. Introgression of "QTL-hotspot" region enhances drought tolerance and grain yield in three elite chickpea cultivars. Plant Genome. 2021;**14**:e20076

[50] Pandey MK, Gangurde SS, Sharma V, Pattanashetti SK, Naidu GK, Faye I, et al. Improved genetic map identified major QTLs for drought tolerance- and iron deficiency tolerance-related traits in groundnut. Genes. 2020;**12**(1):37. DOI: 10.3390/genes12010037

[51] Rane J, Raina SK, Govindasamy V, Bindumadhava H, Hanjagi P, Giri R, et al. Use of Phenomics for differentiation of Mungbean (Vigna radiata L. Wilczek) genotypes varying in growth rates per unit of water. Frontiers in Plant Science. 2021;**12**:692564. DOI: 10.3389/fpls.2021.692564

[52] Zaman-Allah M, Zaidi PH, Samuel T, Jill C, Vinayan MT, Seetharam K. Phenotyping for abiotic stress tolerance in maize: drought stress. 2016. Available from: http://hdl.handle.net/10883/17716

[53] Zhao C, Zhang Y, Du J, Guo X, Wen W, Gu S, et al. Crop phenomics: Current status and perspectives. Frontiers in Plant Science. 2019;**10**:714. DOI: 10.3389/fpls.2019.00714

[54] Cast S, Lobet CG, Cabrera-Bosquet L, Couvreur V, Pradal C, Tardieu F, et

al. Connecting plant phenotyping and modelling communities: Lessons from science mapping and operational perspectives, in silico. Plants. 2022;**4**(1):diac005

[55] Pandey P, Irulappan V, Bagavathiannan MV, Senthil-Kumar M. Impact of combined abiotic and biotic stresses on plant growth and avenues for crop improvement by exploiting physio-morphological traits. Frontiers in Plant Sciences. 2017;**8**:537. DOI: 10.3389/fpls.2017.00537

[56] Naqvi RZ, Siddiqui HA, Mahmood MA, Najeebullah S, Ehsan A, Azhar M, et al. Smart breeding approaches in post-genomics era for developing climate-resilient food crops. Frontiers in Plant Science. 2022;**13**:972164. DOI: 10.3389/fpls.2022.972164

[57] Kumar PKC, Bellundagi A, Krishna H, Mallikarjuna MG, Thimmappa RK, Rai N, et al. Development of bread wheat (*Triticum aestivum* L) variety HD3411 following marker-assisted backcross breeding for drought tolerance. Frontiers in Genetics. 2023;**14**:1046624. DOI: 10.3389/fgene.2023.1046624

Chapter 10

Indonesian Toona Breeding Strategy: Comprehensive Review and the Application Status

Jayusman and Budi Utomo

Abstract

This research was conducted to investigate the need for a breeding strategy for *Toona sinensis* Roem. The initial section examined the assessment of the foundational population derived from community forests during the years 2006–2008. This included an evaluation of genetic diversity, mating system analysis, and the estimation of parameter values for genetic growth. Therefore, the primary aim of *T. sinensis* breeding program was to enhance productivity. To achieve this goal, fundamental requirements were identified for devising a breeding approach for *T. sinensis*. The second section discussed the assessment of six key areas, namely (1) breeding goals, (2) access to fundamental and breeding populations, (3) selection and enhancement, (4) genetic testing, (5) family relationship management, and (6) reproduction. The results showed that a considerable influence was exerted on the efficacy of the selection. Furthermore, the importance of formulating clear and focused objectives was analyzed with an emphasis on one or two specific aims. It was crucial to acquire a comprehensive understanding of reproductive biology, gene activity, and the interplay between genotype and environmental factors. Suggestions for *T. sinensis* included formulation of a breeding strategy, establishment of a dedicated breeding population, creation of seed orchards, distribution of high-quality seeds, and enhancement of productivity within community forests.

Keywords: breeding population, breeding strategy, breeding program, community forest, and *Toona sinensis* Roem

1. Introduction

Indonesian *Toona sinensis* Roem is a species widely planted in community forests because of its wide range of uses. The leaf contains carotenes, amino acids, and vitamins, along with extracts of Surenin, Surenon, and Surenolactone [1–3]. The species exhibits anti-cancer and anti-tumor properties [4], as well as anti-H1N1-Pandemic influenza A virus activity [5]. The organic leaf extract, containing 2-anti-trypanosomal terpenoid, shows a good effect on *Trypanosoma brucei* rhodesiense, with values ranging from 7.18 to 31.25 μg/ml. Moreover, *T. sinensis* can be cultivated on a substantial scale with minimal growth requirements [6]. The fundamental challenge for the development is to achieve the target by 2025 by increasing the yield from 19 to 30 m^3/ha/year and optimizing the economic value of the plant [7]. Increased forest productivity can be achieved through plant breeding [8] and the success of the program is strongly influenced by the selection of appropriate strategy. *T. sinensis* breeding activities in Indonesia are not fully organized and can lead to duplication of activities and inefficient use of resources and time. The synthesis of results on the species is limited and dominated by the aspect of seed treatment and evaluation of seedling development. Meanwhile, the aspect of propagation is still very limited and scattered in many areas. Research institutes, companies, and universities have contributed to the challenge of assessing its current status. Breeding efforts commenced between 2006 and 2011, including the exploration and acquisition of genetic resources from different large islands. This endeavor resulted in the successful collection of 50 populations of *T. sinensis.*

Ex-situ conservation management as the basis for breeding program can be applied to assess species, breeding, and genetic processes. Breeding is conducted on naturalized and native populations, accompanied by testing and selection efforts. The evaluation process should focus on objective criteria such as genotype × environment interaction, testing of advanced generations, combination of mating plan, genetic parameters, and selection of plants. In addition, the traits of interest and the costs associated with breeding are also worth considering with the role of biotechnology in livestock [9]. *T. sinensis* breeding program has progressed through several stages, excluding the selection and testing of advanced generations [10].

Genetic testing through multisite progeny test was carried out in 2011 by testing 100 families per 5 hectares to generate many interactive groups [11]. This information became the basis for the development of *T. sinensis* using non-interacting families depending on breeding site. In line with the establishment of ex-situ conservation base population and the progression of genetic test breeding, it is important to formulate a comprehensive breeding strategy for *T. sinensis*.

2. Evaluation of breeding strategy base elements

Conventional plant breeding largely depends on phenotypic selection and breeder's experience; therefore, the breeding efficiency is low and the predictions are inaccurate [12] so it's perfect breeding strategy needs to be carried out. Breeding program aims to (1) raise the base and breeding population, (2) reproduce the upgraded genetic material to build a superior population, (3) maintain variation and size in the base and breeding populations, and (4) achieve economic goals.

2.1 Existing base principles

The success of breeding program is largely determined by the ability to meet the 9 (nine) base principles [13] and the evaluation results are as follows:

2.1.1 Evaluation of T. sinensis breeding program

The evaluation of breeding program for *T. sinensis* has successfully achieved the following: (1) establishment of the first generation of *T. sinensis* breeding to enhance productivity through the improvement of height, stem diameter, and volume, (2) mastery of base biology on aspects of phenology and reproduction, (3) identification of genetic material sources from community forest stands, strategically selected from diverse populations and geographic origins, (4) exploration of alternative species, *Toona sureni*

Merr. [8, 14] known for its woodworking properties and in line with the Meliaceae family, (5) availability of seedling sources for progeny testing, (6) implementation of genetic testing through multi-site progeny testing, (7) execution of field operations and research activities in several test plots, (8) strong organizational commitment with research and development tasks, and (9) adherence to a program that combines economic value traits with the management of a broad genetic base, facilitating adaptation and development. Compliance with these base principles shows their level of fulfillment, providing a basis for establishing a breeding program for *T. sinensis*. The objectives of forestry plant breeding programs have been widely reported among others for forest tree improvement [15], evaluation of Genetic diversity and breeding strategies of Azadirachta indica plants [16]. Establishing domestication strategies for forest trees [17] and for population genetics studies in forest tree improvement [18].

2.1.2 Evaluation of breeding program compliance

Objectives are based on the greatest long-term genetic gain achieved through effective selection, based on large and variable-sized populations with relationship control in future breeding generations. The development of base and breeding populations can use progeny test plots supported by moderate genetic diversity and hybridization patterns of the parent plants. The availability of the base population facilitates the selection of broodstock to establish progeny trial orchards for F-1 seed production and vegetative propagation to support clonal production and cultivation. The management of *T. sinensis* base and breeding populations to maintain genetic diversity can be improved through the transmission of exogenous populations and inter-families (breeding) to produce superior hybrids (*T. sinensis* Roem > < *T. sureni* Merr). The research implementation is sustainable, and meeting the set goals in terms of time and budget efficiency is always evaluated and improved.

2.1.3 Short-term goals of Indonesian T. sinensis breeding program

Quality seeds can be produced by converting the first-generation progeny test. Selection in progeny test plots is performed within families (single tree plots) and the representativeness can always be maintained quantitatively to manage the genetic base of the original population. The development of second-generation progeny testing is optimized through the transmission of genetic material from populations outside first-generation testing. These efforts are consistent with the view that the success of breeding program is determined by two important aspects, namely (1) the rapid acquisition of the desired product and (2) long-term retention to prepare a population with a broad base of genetic diversity for next-generation selection program. Even though the concept is not emphasized in short-term program, it is important in the long term [19].

2.1.4 Constraints on the genetic material in the base population

Restricted genetic material in the base population should be predicted by transferring from an outside population or crossing to produce second-generation breeding (**Figure 1**). The need for selectable genetic material is essential in the cycle of forestry tree breeding program [13, 20]. Cycle diagram of a breeding program through inbreeding and transmission activities has the potential to increase genetic diversity in the base population of the next generation [20]. Maintaining and enhancing genetic diversity is an important foundation of the subsequent cycle of breeding program (**Figure 2**).

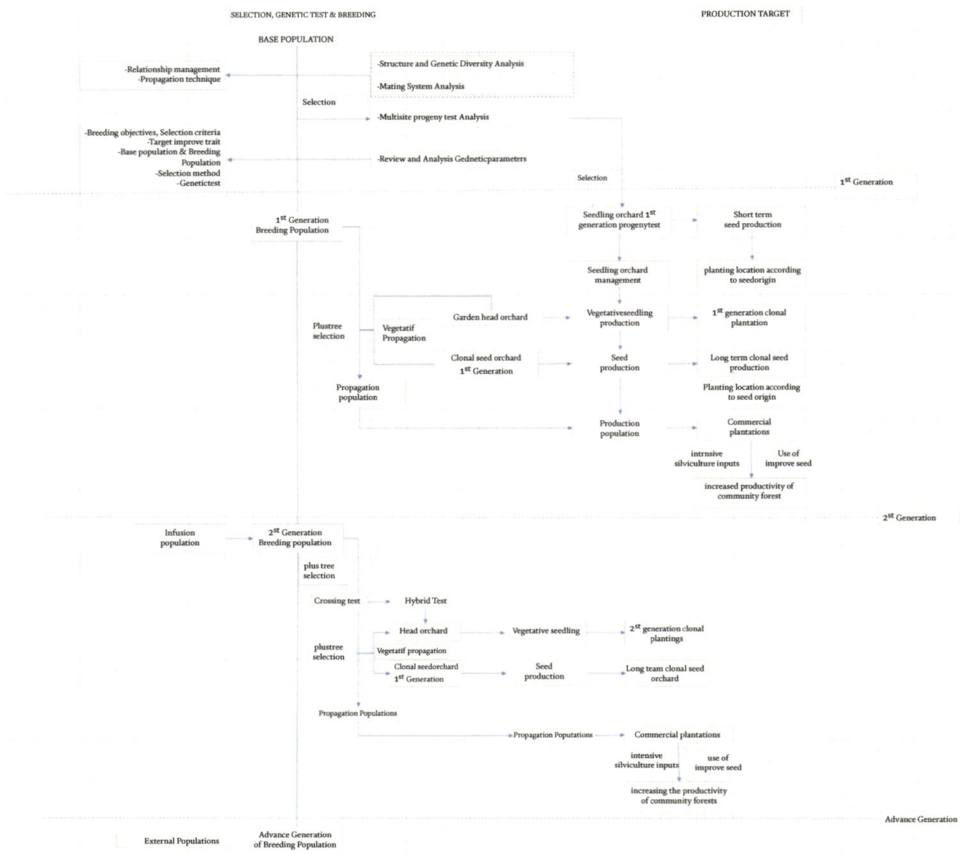

Figure 1.
T. sinensis Roem. Breeding strategy based on all population levels connected in the line (base population, breeding population, propagation population and production population) and on generation I and generation II.

Toona sinensis Forest Community *Toona sinensis* Mother Tree

Figure 2.
T. sinensis community forest plantation and mother tree.

3. Breeding strategy

The plan for the attainment of desired objectives within breeding program, which considers all the conditions, is referred to as breeding strategy. The execution is termed breeding, while the use of biological processes in propagation such as grafting, controlled pollination, and field testing, is recognized as a breeding strategy. This strategy determines the most effective system for managing different areas of the program considering the constraints of time and resources. Furthermore, [20] explained that the process included (1) specific design, (2) time-matched, and (3) logistical application of all plant propagation components. These are selection screening, testing, breeding maintenance, breeding population development, commercial release, transmission of new genetic resources, and conservation and research efforts.

Evaluation of breeding strategy to be established concerning the aspects of benefits and constraints faced has been carried out on *Azadiracta indica* [18], *Triticum* spp. [21], organic plants [16], *Pinus pinaster* and *Eucalyptus* [22], *Tectona* and *Eucalyptus* [23], as well as drought tolerance mechanisms in plant [24]. Many breeding strategy evaluations are reported using different methods. These include references to the magnitude of genetic gain [18], accomplished through simulation program grounded in gene action [25]. Additionally, there is a use of simulations to evaluate the gains and losses incurred by the applied breeding strategy [26]. The use of variance components to assess the ramifications of alterations in field trial is conducted through different factors such as the number of genotypes, years, locations, and replications [23]. Modeling and simulation of plant breeding strategies has also been published [13]. The examination of alternative strategy representing intensive tree breeding is also conducted [26].

The research focuses on evaluating the compliance of the required elements of *T. sinensis* breeding strategy. This includes (1) breeding objectives, (2) availability of base and breeding populations, (3) selection and genetic acquisition, (4) genetic testing, (5) relationship management, and (6) reproduction. The evaluation of breeding strategy compliance is the basis for determining *T. sinensis* by considering existing conditions to ensure the achievement of predetermined targets.

3.1 Breeding objectives review

Breeding objectives of *T. sinensis* are focused on increasing plant productivity through the observation of important growth traits selected, such as height growth, stem diameter, stem straightness, and volume. The criteria used to increase productivity must be accompanied by improvements in stem straightness for woodworking purposes. Meanwhile, the selection of parent trees obtains plus trees used to build breeding and propagation populations. The initial criteria set for plus trees are to have desirable traits such as maximum stem straightness, fast growth, cylindrical stem, large diameter, small branching with a horizontal angle, pest and disease resistance, and other specific traits according to breeding objectives.

Selection criteria for woodworking can be adjusted specifically for growth traits having a major influence on tree quality. Information based on the evaluation of growth genetic parameters (height, diameter, stem straightness, and volume) can be used as a basis for selection activities. The strategy used to obtain the greatest genetic gain have been carried out through direct, and indirect selection based on multi-location tests. Evaluation of genetic parameters produces information to fulfill the elements of breeding objectives in the preparation of strategy.

3.2 Base and breeding populations

The availability of base and breeding populations has been fulfilled by the progeny test plots, which also fulfill the element of genetic testing. The analysis in the form of progeny tests in advanced-generation program can be used as base population. The presence of genetic material in the progeny test plots reduces operational costs for efficiency purposes because there is no need to repeat activities in the field (**Figure 3**).

The selection for the subsequent generation of breeders is made from base population. In the first generation, *T. sinensis* population was primarily from unreforested community forests. In the second generation, the base population can be a test plant genetically derived from selected parents of the previous. Breeding population is a group of individuals selected from the base and become the seniors of the next generation. The mating system in a population reproduces additional genetic variation to achieve a continuous increase in the frequency of dominant genes in the base.

The plan developed to achieve the desired goals for the existing conditions is called breeding strategy, and the implementation process is known as breeding method. The biological processes used in propagation, such as grafting, controlled pollination, and field testing are called breeding strategy. The strategy aims to determine the most effective system for managing different areas of the program considering the constraints of time and available resources. Meanwhile, [20] explains that breeding strategy includes (1) specific design, (2) time-matched, and (3) logistical application of all plant propagation components. These include selection screening, testing, breeding maintenance, population development, commercial release, transmission of new genetic resources, and conservation and research efforts.

Evaluation of breeding strategy concerning the aspects of benefits and constraints has been carried out on *Azadiracta indica* [18], *Triticum* spp. [23], organic crops [9], *Pinus pinaster* and *Eucalyptus* [22], *Pinus radiata* [13, 27], *Tectona* and *Eucalyptus* [16], and drought tolerance mechanisms in crop plants [24]. Many breeding strategy evaluations are reported using different approaches referring to the magnitude of genetic gain [18] through simulation program. This is based on gene action [25], using simulations to assess the gains and losses of breeding strategy applied [26] and

Toona sinensis ex-situ conservation   Toona sinensis Progeny Test

Figure 3.
Ex-situ conservation and progeny test Indonesia Toona sinensis.

variance components to evaluate the changes in field trial selection strategy. This is achieved by varying the number of genotypes, years, locations, and replications [23] and evaluating alternative strategy representing intensive tree breeding [8].

The present research focuses on evaluating the compliance of the required elements of *T. sinensis* breeding strategy. This includes (1) breeding objectives, (2) availability of base and breeding populations, (3) selection and genetic acquisition, (4) genetic testing, (5) relationship management and (6) reproduction. The compliance of breeding strategy is the basis for determining *T. sinensis* by considering all existing conditions to ensure the achievement of predetermined targets.

3.3 Breeding objectives review

Breeding objectives of *T. sinensis* Roem are focused on increasing plant productivity through the observation of important growth traits, including height growth, stem diameter, stem straightness, and volume. The selection criteria used to increase productivity must be accompanied by improvements in stem straightness for woodworking purposes. Selection of parent trees is aimed at obtaining plus trees to build breeding and propagation populations. The initial criteria are to have desirable traits such as maximum stem straightness, fast growth, large diameter, small branching with a horizontal angle, and pest and disease resistance, according to breeding objectives.

Selection criteria for *T. sinensis* for woodworking can be adjusted especially for growth traits with a major influence on tree quality. Information based on the evaluation of growth genetic parameters (height, diameter, stem straightness, and volume) can be used as a basis for selection activities. The strategy to obtain the greatest genetic gain have been carried out through indirect, and direct selection based on multilocation tests. Evaluation of genetic parameters produces information to fulfill the elements of breeding objectives required many breeding strategy evaluations have been published including simulation of gene action and dryland environment effects [28], based on partial pedigree reconstruction through simulation [29] and Evaluation of testing strategies for plant breeding field trials [23] in the preparation of breeding strategy.

3.4 Base population and breeding population

The availability of base and breeding populations has been fulfilled by the progeny test plots in line with the element of genetic testing. The analysis of breeding population in the form of progeny tests in advanced-generation program can be used as the base population. The presence of genetic material reduces operational costs for efficiency purposes because there is no need to repeat activities in the field.

The base population is used for the selection of the next generation of breeders. In the first generation, *T. sinensis* populations were primarily obtained from community forests. In the next generation, the base population can be a test plant genetically derived from selected parent plants of the previous generation. A breeding population is a group of individuals selected and become the seniors of the next generation. Breeding population passes *T. sinensis* species from one generation to another. The mating system in a population reproduces additional genetic variation to achieve a continuous increase in the frequency of dominant genes. Breeding populations typically include 200 or more selected plants.

The development of progeny trials in multiple sites as breeding populations provides *T. sinensis* from one generation to another. Selection strategy use genetic variation to achieve a continuous increase in the frequency of the best genes within the families. The process of selecting appropriate genetic resources for the establishment of breeding populations includes the utilization of a population management strategy for selection and breeding functions. To generate a breeding population, a minimum of 20–30 individuals is necessary from the population to produce seeds or vegetative seedlings for the establishment of a commercial plant. The opportunity to obtain individuals from progeny test plots is available by selecting within the family using weights adjusted for preferred traits. Multisite progeny test plots can function as base populations and propagate to selectively advance progeny. A breeding population consists of several individuals (20–30) of plants obtained through intensive selection to produce seeds or vegetative sprouts for propagation of commercial plants. Furthermore, it can be in the form of a seed or pruned orchard and the genetic material is used for commercial cultivation. Based on the evaluation, breeding and the production population of *T. sinensis* are not available, hence the concept is proposed as a target for development in the next selection phase.

3.5 Selection and genetic enhancement

Many breeding programs create multiple traits and these activities require information on several characteristics in the selection process. Based on the results, selection using the index method is a widely developed system for selecting many traits. The use of the combined selection index makes it possible to assign a score to each individual. In addition, the economics of each trait should be considered when constructing an index of selection. The general selection process includes (1) identification of traits to be evaluated, (2) observation of preferred traits in the parent and breeding populations, (3) data analysis, (4) calculation of the indices for each plant, (5) ranking the best candidate trees, (6) analyzing the genetic gain for each trait, and (7) caring for the selected plants [20]. Genetic selection and augmentation are important steps in breeding cycle to determine the success of breeding strategy.

The selection was tested in *T. sinensis* progeny evaluation using the weights of all traits and values to compile the index. Based on the selection index, the methods of direct, indirect, and combined selection were analyzed. The analysis shows that direct selection has the greatest genetic benefit compared to indirect and combined selection. The implementation of the direct selection on *T. sinensis* completed the selection elements in the strategy (**Table 1**).

Traits	**Locations**					
	Candiroto			**Ciamis**		
	Direct Selection	**Indirect Selection**	**Combined Selection**	**Direct Selection**	**Indirect Selection**	**Combined Selection**
Tree Height (m)	12.36	0.39	9.48	1.05	0.10	−0.14
Diameter (cm)	12.23	0.004	9.09	0.77	0.01	−0.01
Stem Form	1.74	0.02	1.43	6.23	0.01	4.85
Tree Volume ($\times 10^{-3}$ m^3)	28.39	0.02	21.19	2.50	0.01	−0.42

Source: [30].

Table 1.
Estimation results of genetic gain based on family selection using direct, indirect, and combined selection methods.

Combined ANOVA generates values based on a single trait, while selection is made using multiple traits. Therefore, indirect selection remains important in generating information on changes in the ranking of families and becomes the basis for determining the families to join breeding program. The genetic gains in this multisite trial were diverse, with the greatest value produced by direct selection, while indirect and combined selection produced lower values for all four tested traits.

3.6 Genetic testing

Progeny testing selects parents based on the performance of the seedling. This selection is useful in providing information on the genetic thinning of seed orchards. The test is an excellent strategy because the mean of the progeny is an estimate of the heritability. Analysis of genetic parameters shows that the multisite progeny test can be used for single-site and multisite analyses. The ability to view heritability, genetic correlation, and amplification information from multisite testing can be confirmed based on single-site analysis. The establishment of progeny trials at two sites fulfilled the genetic testing elements of *T. sinensis* breeding requirements.

3.7 Relationship management

The element of relationship management was met by analysis of genetic structure and diversity as well as by analysis of reproductive systems. Genetic diversity is an important fundamental principle in terms of ethical aspects when managing stands, ecosystems, and landscapes [17]. Connection between species diversity and genetic diversity [31]. The distribution is a prerequisite for adaptation that determines the long-term stability of individuals, species, and entire ecosystems. Genetic diversity should be assessed in long-term collections, breeding populations, seed orchards, or production populations (**Figure 4**) [20]. Genetic variation is a prerequisite for future evolution and we stress that gene conservation programs should provide opportunities for future evolution [33].

The use of a genetic basis offers the opportunity to select the superior traits necessary for the wood development. Some of the characteristics required for woodworking include a straight stem, optimum height without branches, minimal knotting

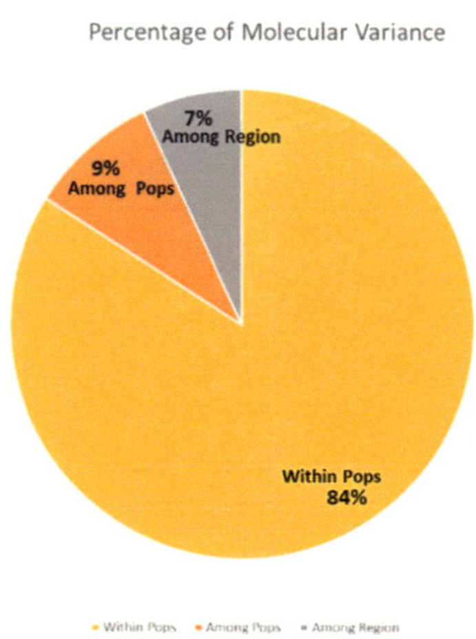

Figure 4.
Percentage of molecular Variance T. sinensis mother tree [32].

Species	tm	tm-ts	References
Toona sinensis	0.94	017	[11]
Melia azedarach	1.00	0.00	[34]
Swietenia macrophylla	0.96	0.14	[35]
Carapa prucera	0.78	0.01	[22]
Cedrela odorata	1.06	0.16	[36]
Carapa guinensis	0.86	0.13	[37]
Azadirachta indica	0.90	0.04	[38]

Source: [30].

Table 2.
Comparison of outcrossing value at many of the loci (tm) and outcrossing value at one locus (ts) in several plant species.

defects, good machining properties, and beautiful patterns. *T. sinensis* has a constant density suitable for molding materials, and the characteristics of processing into sawn products, veneer, and furniture are not difficult.

Breeders must understand the extent of inbreeding within the populations. The discovery of the important role of *T. sinensis* outcrossing has strategic implications for the established experimental plots of progeny and for reinforcing the selection strategy (**Table 2**).

The out-crossing information in the navigation system shows that *T. sinensis* has superior hybrid characteristics used to maintain genetic diversity for the next generation. Seed management is expected to avoid the mixing of families with the same relevant population. The impact of mixed breeding on *T. sinensis* population requires alternative strategy for the program. The main implication is to prepare opportunities for the use of selection schemes including similar populations, hybridization, or repeated selection to achieve recombination. This is genetically engineered by random crossbreeding to achieve sustainable production and genetic progress.

3.8 Propagation method

The reproduction of genetic material can be achieved by propagation and vegetative methods. Vegetative method is selected when (1) propagation of mature individuals is conducted and (2) propagation of immature individuals is performed. Meanwhile, propagation is selected when (1) seeds are highly produced at a relatively young age and (2) vegetative method proves unfeasible.

Assessing the suitability of proficiency in reproductive propagation for *T. sinensis* is largely mastered from aspects of plant phenology, seed treatment, and generational propagation. Breeding strategy of *T. sinensis* is largely respected and can serve as the basis for defining the concept. Compliance with the requirements of each component must be ensured, including the anticipation of certain changes due to technical and management factors.

4. Examining the determinants of the selection strategy

Breeding strategy of *T. sinensis* is inseparable from the many factors that determine the success of the concept, including (1) breeding goal is not too complicated and should be limited to 1 or 2, and (2) the controllability of reproductive biology, (3) mastering information on the role of genes (gene activity) including plus and non-additive traits, (4) the existence of genotype-environment interactions, and (5) potential future. The phenomenon of hybrid superiority can be related to heterozygosity or complementarity. Heterozygousness (hybrid viability) is the result of the role of non-resonant genes while complementarity is the role of additive genes and related to how two or more traits complement each other in a particular environment. Other aspects that play an important role in the success of a breeding strategy are the appropriateness of budgets for breeding activities and production populations as well as the availability of (1) genetic resources, (2) human resources, (3) information, and (4) infrastructure.

The initial assessment was predicated on the factors influencing the efficacy of breeding strategy for *T. sinensis*. A majority of these factors have been satisfactorily addressed, including the identification of the primary objective. The commencement of breeding initiative centers on augmenting yield. Moreover, the scrutiny of the Reproductive Biology control has yielded valuable insights, accompanied by an analysis of genetic and environmental interactions. The key to the success of urgently identified and implemented breeding strategy is the aspect of understanding the role of genes and aspects of their potential hybrid ability. The advantage has been emphasized by [19] that hybrids can generate gene combinations while paying attention to the diversity of the old hybrids. The potential of *T. sinensis* hybrid is quite promising, which is supported by the existence of two species, growing naturally, namely *T. sinensis* Roem and *T. sureni* Merr.

The first limitation of the information that deserves attention is the degree of pollination and fertilization compatibility produced by the two species in heterozygous hybrids. The steps begin with controlling pollination of two different species and are aided by easy vegetative propagation of superior hybrid clones to develop the operation of Eucalyptus and Populus hybrid programs [20]. F1 hybrids, resulting from crossing between two different species, have become commercial hybrids. F2 hybrids (resulting from F1 hybrids), and F1 hybrids are hybrids of F1 with F1 hybrids one of the parent species. *T. sinensis* hybrid development program is part of the strategy to increase productivity in the future.

4.1 Selection strategy

The propagation of *T. sinensis* should be based on an appropriate breeding strategy, supported by a high degree of technological mastery and the application of appropriate breding. Selection plays an important role in determining genetic gain and the process in progeny trials requires a great deal of additional information for the selection to avoid loss of genetic potential. The risk of genetic loss is greater when breeding to improve multi-trait propagation at multiple breeding stages. Furthermore, the selection of multi-trait varieties is strongly influenced by the behavior, traits, and uses of the selected plants. Efforts to obtain suitable selection criteria for multiple preferred traits were particularly important for *T. sinensis* for furniture. 1st.

Crossbreeding for the simultaneous improvement of multiple traits depends on whether each trait is controlled by a single or group of genes. Independent characters can be selected consecutively or simultaneously. Traits that are negatively correlated must be carefully estimated because an increase equates to a decrease in the level of another. The use of selection weights and indices is consistent with the recommendation of [38] that the use of selection indices to improve multiple traits is a good choice.

The complexity of using a single feature as the basis for selection and the bottlenecks at the operational stage can be reduced by the selection index.

The preparation of breeding populations requires information that evaluates the genetic parameters. Base information includes genetic value, correlation, and gain. Heritability is the rate of variation in a population caused by genetic differences between individuals. Therefore, this ratio indicates the extent to which parental traits are passed on to offspring. The level of heredity should also be considered an important factor in the success of selection for genetic traits. The genetic value varies depending on the type and age of the tree. Since the concept is age-dependent, early selection can be used for efficiency purposes. Based on heritability at a location, all traits, such as height, diameter, straight body, and volume, showed higher values than between sites. The genetic values are the basis for calculating the gain using direct, combined, and indirect selections. The results are used as the basis to develop seed production for site-specific selection (propagation) to avoid loss of genetic gain.

4.2 Relationship management strategy

The genetic relationship management concept existing in the base population must be managed, maintained, and maximized in terms of genetic diversity. The selection of parental plants and the construction of breeding populations that will use information on the genetic distance to create a basis4 remains wide-ranging [34, 36]. Every plant breeder must innovate to create new variations and improve breeding to meet target needs [10]. This previous research is strongly supported by the mating pattern of *T. sinensis* prone to dominant outcrossing [25], where the parent plants resemble each other. Compliance is assessed with elements of *T. sinensis*, including relationship management and methods, making recommendations for the management of seed orchards.

4.3 Seed production strategy

A short-term strategy could be to use seeds from progeny test plots converted into first-generation seed orchards. Efforts to convert progeny test plots into seed orchards are strongly supported by information on the genetic diversity of Mother plants (0.304). The results of the evaluation of mating system in the progeny test plots showed that *T. sinensis* tended to engage in outcrossing (93.6%), inbreeding (17.2%) and selfing (6.1%) [30, 32] respectively. The species with high degree of hybridization should be able to maintain their genetic diversity [30, 32].

The requirements to convert the progeny trial pots to the nursery were evaluated and finalized including area, genetic diversity, species mating system, flowering, and fruiting time. The requirement for genetic diversity ensures and takes advantage of the population in adapting to changes in the environment. The conversion of *T. sinensis* progeny test plots into first-generation seedling orchards offers an advantage as a short-term source of seed through propagation that leaves 1 plant per plot (one-tree plot) in cut-down trees with suboptimal growth rates and lower yields. Cutting trees as part of a breeding operation creates a wider distance for pollinators to transfer pollen from one tree to another for increased outcrossing value. Furthermore, flowering stimulation was applied using existing technology to achieve fruiting information on the genetic diversity of parent plants, which was (H = 0.304) and 0.024 higher than *T. sinensis* population in West Java [26] and equivalent to those in China 0.333 [39].

There are two alternatives considering the role of genetic diversity in the management of progeny test populations. The first includes progeny test populations as a means of transmitting genetic material for the construction of breeding populations. This is achieved through stringent selection criteria applied to selected trees while ensuring that the requisite woodworking traits are present within the progeny test population. The second option includes leveraging genetic diversity information as the foundation for batch management during progeny testing within the first-generation seedling orchard lineage, with a focus on short-term seed production. The overarching objective is to facilitate the transmission of genetic diversity and yield-related information from parental plants to their progeny during the testing process.

4.4 Population management strategy

The strategy of establishing a foundational population with a restricted genetic pool comprised of 100 families from 10 diverse geographical origins. This necessitates the incorporation of genetic material from external sources into breeding population in response to external influences. The number of families participating in the base population is greater than 200 families that are representative of the population in terms of distribution and range. Information on the distribution of genetic diversity of *T. sinensis* is more abundant within populations (84%) and between geographical sources 9 and 7%. The genetic resource naming is to gather more resources from geographical sources [32].

The enhancement of genetic diversity of breeding populations beyond transmission into external populations can be achieved through crossing families with large genetic distances as well as hybridization between *T. sinensis* Roem and *T. Sureni* Merr species to create superior hybrids. More crossbreeding increases the likelihood of multiple combinations and increases the genetic diversity of breeding population [20]. According to the concept of the selection cycle, increasing the genetic diversity of breeding population and genetic testing are the main factors that are considered complete.

Research on the structure and genetic diversity of *T. sinensis* progeny showed that the trend maintained the original genetic diversity. However, research is needed to ensure that the original genetic diversity is maintained to ensure the stability of the testing in the next generation of livestock. The decline in genetic diversity may be due to a reduction in the number of effective populations engaged in next-generation selection. Information on the genetic diversity of *T. sinensis* carries two recommendations, namely (1) assessing the expected genetic gains for important traits and (2) evaluating the mating system of *T. sinensis* population in the next generation of breeding.

4.5 High-yielding hybrid, clonal, and hybrid strategy

Breeding strategy for *T. sinensis* should focus on seed manipulation for generational and vegetative propagation methods of plus plants to support clonal production and plantation development in the nucleus. In first-generation breeding, clonal propagation relies on the selected plant and the outcomes. However, second-generation breeding places a greater emphasis on a combination of plants and progeny selection testing, including System II with the integration of genetic material from hybrid plants. The last major section on new technologies highlights on of which is the increasing importance of hybrid breeding in tree improvement programs [40].

One of the options of the cloning method is to optimize the inheritance of maximum characters that are superior to their descendants. The use of cloning in addition to genetic resources and plant growth can be developed to clone superior hybrids. The search for superior hybrids can increase the genetic diversity of populations and potentially create more choices. Indonesia has natural populations of Toona sinensis & Toona sureni that have potential for hybrid programs. Initial identification of physical & anatomical properties of both Toona species has also been carried out [14]. The creation of hybrids promises variety and new breeding opportunities for traits produced by hybrids.

5. Family development strategy

Families with the lowest distance values have the most stable characteristics in the multisite test because the rankings do not differ at each or combined site. A maximum standard deviation value of 10 was set as a reference to identify the least interacting and most stable families in the preliminary period. Furthermore, the large number of families interacting in the multisite trial may be due to the sensitivity of families to environmental fluctuations. Controlling the number of families interacting in *Pinus pinaster* [41] was performed by excluding from the subsequent selection program. However, applying this approach to testing *T. sinensis* is not entirely feasible since only 58 of the 100 families were included in the cross-site test. Removal of all interacting families results in a decrease in genetic potential due to the loss of valuable material.

The preservation of families with lower differential values can be considered as a measure of the conservation of genetic material. The identification of higher-order stable families is important for the development of late-stage *T. sinensis*. Considering the best ranking from 1 to 20 (35% of the 58 families tested multi-site), the 6 families with the highest rank and the lowest interaction value were selected [30].

6. Quality seed distribution strategy

The multisite trial analysis shows that surname and site have a very strong influence and the sensitivity of the seed produced will be affected by each planting site. White et al. [20] mentioned that one of breeding strategy for logistics implementation was to manage the distribution of commercial products. Therefore, *T. sinensis* seed distribution for community forest development needs attention. Seeds should be distributed in areas with similar soil and climatic conditions as the seed source. To enhance the productivity of community forests with the use of high-quality seeds during the selection phase, it is important to implement intensive silvicultural practices. High-yielding seeds need to be supported with the right planting site to realize their genetic potential. Most improved materials currently deployed are seed crops from first-generation phenotypic or tested seed orchards, which offer 10–25% gains in yield depending on the selection intensity of parent trees [42].

7. *T. sinensis* community forest development strategy

Based on *T. sinensis* breeding strategy in **Figure 1**, short-term support can be achieved using seed production from the nursery of generation trial seedlings first-grade descendants. This reduces reliance on seed use dependent on sources from identified seedlings. The quality hierarchy of the nursery is determined to be lower than the nursery. Therefore, a short-term program of plant genetic improvement is quite strategic to achieve every *T. sinensis* yielding 2–3 kg with the number reaching 40,000–60,000 seeds/kg.

The results of examining aspects of the interaction show that family and habitat have a strong influence, hence, it is recommended to distribute seeds by adjusting the soil and climate conditions of the community forestry development area. The development of seed resources based on planning can be considered as adjusting the distribution of community forest culture. Incorporating forest development planning reduces the chance of inappropriate seed transfer and can optimize productivity and growth.

The strategy in **Figure 1** exhibit the most favorable ratings and show reduced interaction. Even though the availability of known parent and family plants with the highest-ranking advantages and minimal interaction remains constrained, efforts can be directed toward optimizing their use to bolster clonal propagation, establish cloned seed orchards, and foster the growth of asexual plants. The cloning strategy has been developed through vegetative propagation with promising results. Propagation of *T. sinensis* gene by vegetative method is under control [43] and the advantages are well-known to breeders. However, the application requires serious economic consideration, availability of human resources, technology, and budgetary support.

For the realization of long-term benefits, it is important to formulate activities aimed at preserving sufficient genetic diversity to anticipate alterations in the reproductive system within the subsequent generation. Knowledge of mating systems can prove invaluable in enhancing established selection methods. A range of strategy may be evaluated and deployed to broaden the genetic foundation of breeding populations for enduring programs, such as crossbreeding to yield superior hybrid offspring.

The high-quality seeds produced from the nursery must be tested against the conditions of the community forest which has a mixed silviculture pattern with irregular tree spacing and spread over a small planting area. Efforts to determine the superior genotypes suitable for community forest soil conditions can be achieved through seed testing under environmental conditions appropriate to the growing site.

Due to the elevated sensitivity in the tested *T. sinensis* families, the optimal selection approach uses direct selection rather than conducting further testing. Consequently, the transfer strategy for the cultivation should prioritize the use of high-quality seeds from nurseries within community forest development sites possessing soil conditions and climates.

8. Conclusions

In conclusion, *T. sinensis* breeding strategy was improved when establishing breeding populations using previously collected base population data. The establishment of seed orchards was based on the location of cultivation development by restricting the distribution of excellent families to designated breeding areas. Furthermore, the development of species community forest was prioritized using seed sources with soil and agro-climate suitable for the development area. The key feature of any successful breeding strategy was providing flexibility to make changes since information formed part of new technology.

Acknowledgements

The authors are grateful to Prof. Muhammad Na'iem for the valuable suggestions for this research and to Dr. Arif Nirsatmanto for the support and valuable suggestions during the writing process and discussion. The authors are also grateful to the Ministry of Environment and Forestry of Indonesia for providing scientific support on Breeding Program of Indonesian *Toona*.

Conflict of interest

The authors declare no conflict of interest.

Author details

Jayusman[1*] and Budi Utomo[2]

1 Research Center for Plant Conservation, Botanic Garden and Forestry, National Research and Inovation Agency (BRIN), Bogor, Indonesia

2 Faculty of Forestry Silviculture Department, Universitas Sumatera Utara, Medan, North Sumatra, Indonesia

*Address all correspondence to: jayu001@brin.go.id

References

[1] Edmonds JM, Staniforth M. *Toona sinensis* (Meliaceae). Curtis' Botanical Magazine. New York, United States of America: Wiley; 1998;**15**(3):186-193

[2] Silva MFGF, Agostinho SMM, Paula JR, Neto JO, Gamboa IC, Filho ER, et al. Chemistry of Toona ciliata and Cedrela odorata graft (Meliaceae). Chemosystematic and ecological siginificance. Pure and Applied Chemistry. 1999;**71**(6):1083-1087

[3] Chang HC. *Toona sinensis* Roem. American Journal of Chinese Medicine. 2002;**30**(2 & 3):307-314

[4] Yuan SSF, Shiang YC. The fractionated *Toona sinensis* leaf extract induces apoptosis of human ovarian cancer cells and inhibits tumor growth in a murine xenograft model. Gynecologic Oncology. 2006;**102**(2):309-314

[5] You HY, Chen CJ, Eng HL, Liao PL, Huang ST. The effectiveness and mechanism of *Toona sinensis* extract inhibit attachment of pandemic influenza a (H1N1) virus. Evidence-based Complementary and Alternative Medicine. 2013;**2013**:479718, 12 pages. DOI: 10.1155/2013/479718

[6] Lemmens RHMJ, Soerianegara I. Wong WC. Plant Resources of South East Asia, Prosea. Timber trees: Minor commercial timbers. In: Plant Resources of South East Asia (PROSEA). Vol. 5. No. 1. Bogor; 1995

[7] Forestry Research and Development Agency-FORDA. Road Map Research and Development of Forestry. Jakarta: Indonesian Ministry of Environment and Forestry; 2009. 16 p

[8] Rosvall O, Mullin TJ. Introduction to breeding strategies and evaluation of alternatives. In: Best Practice for Tree Breeding in Europe. Sävar, Sweden: The Forestry Research Institute of Sweden; 2013. pp. 7-27

[9] Nuijten E, Messmer M, van Bueren EL. Concepts and strategies of organic plant breeding in light of novel breeding techniques. Sustainability. 2016;**9**(1):18

[10] Namkoong G, Barnes R, Burley J. A Philosophy of Breeding Strategy for Tropical Forest Trees. UK: Commenwealth Forestry Institute, University of Oxford; 1980

[11] Jayusman J, Na'iem M, Indrioko S, Hardianto EB, Nurcahyaningsih ILG. Out crossing value estimation in *Toona sinensis* Roem based on RAPD markers. Journal Penelitian Kehutanan FALOAK. 2018;**2**(1):13-28

[12] Wang J. Modelling and simulation of plant breeding strategies. In: Plant Breeding. China: Chinese Academy of Agricultural Sciences (CAAS); 2012. pp. 19-40

[13] Dungey HS, Brawner JT, Burger F, Carson M, Henson M, Jefferson P, et al. A new breeding strategy for Pinus radiata in New Zealand and New South Wales. Silvae Genetica. 2009;**58**(1-2):28-38

[14] Jayusman J, Hakim L. Comparison of the wood anatomy and fibers derived from Indonesian *Toona sinensis* Roem. And Toona sureni Merr. BioResources. 2021;**16**(3):4769-4779

[15] Kedharnath S. Forest tree improvement in India. Proceedings: Plant Sciences. 1984;**93**(3):401-412

[16] Kundu SK, Luukkanen O. Genetic diversity and breeding strategies of the

neem (Azadirachta indica). In: XII World Forestry Congress. Sävar, Sweden: The Forestry Research Institute of Sweden; 2003

[17] Libby WJ. Domestication strategies for forest trees. Canadian Journal of Forest Research. 2013;**3**:265-276

[18] Muona O. Population genetics in forest tree improvement. In: Brown AHD, Clegg MT, Kahler AL, Weir BS, editors. Plant Population Genetics, Breeding and Genetic Resources. Sunderland, MA, USA: Sinauer Press; 1973. pp. 282-298

[19] Zobel BJ, Talbert J. Applied Forest Tree Improvement. New York, United States of America: Wiley; 1984. 505 p

[20] White TL, Adam WT, Neale DB. Forest genetics. Chapter 13. In: Phenotypic Mass Selection-Genetic Gain, Choice of Traits and Indirect Respon. United Kingdom: CABI Publishing; 2009. pp. 329-354

[21] Hidayat Y. Perkembangan Bungan dan Buah Pada tegakan benih surian (*Toona sinensis* Roem). Jurnal Agrikultural. 2010;**21**(1):13-20

[22] Doligez A, Joly HI. Mating system of Carapa Procera (Meliaceae) In the French Guiana tropical Forest. American Journal of Botany. Costa Rica: Tropical Agricultural Research, and Higher Education Center; 1997;**84**(4):461-470

[23] Arief VN, DeLacy IH, Crossa J, Payne T, Singh T, Braun R, et al. Evaluating testing strategies for plant breeding field trials: Redesigning a CIMMYT international wheat nursery. Crop Science. 2015;**55**(1):164-177

[24] Spitters CJ, Schapendonk AHCM. Evaluation of breeding strategies for drought tolerance in potato by means of crop growth simulation. In: Genetic Aspects of Plant Mineral Nutrition. Dordrecht: Springer; 1990. pp. 151-161

[25] De Campos T, Da Cunha MO, De Sousa ADB, Teixeira RB, Raposo A, Sebbenn AM, et al. Mating system parameters in a high density population of andirobas in the Amazon forest. Pesquisa Agropecuaria Brasileira. 2013;**48**(5):504509

[26] Hidayat H, Siregar IZ. Preliminary Evaluation On Genetic Variation Of Two Year Old Surian (*Toona sinensis* ROEM) Progeny Test Assessed By RAPD MARKER. 2011. Available from: http://library.forda-mof.org/libforda/files/Proceeding/20INAFOR/2020-11.pdf

[27] Burdon RD. Breeding radiata pine-historical overview. New Zealand Journal of Forestry. 2008;**52**:4

[28] Chapman S, Cooper M, Podlich D. Hammer G: Evaluating plant breeding strategies by simulating gene action and dryland environment effects. Agronomy Journal. 2003;**95**(1):99-113

[29] Bouffier L, Klápště J, Suontama M, Dungey HS, Mullin TJ. Evaluation of forest tree breeding strategies based on partial pedigree reconstruction through simulations: Pinus pinaster and Eucalyptus nitens as case studies. Canadian Journal of Forest Research. 2019;**49**(12):1504-1515

[30] Jayusman. Analisis Parameter Genetik Dan Pemanfaatannya Untuk Strategi Pemuliaan Surian (*Toona sinensis* Roem.). Disertasi Program Studi Ilmu Kehutanan. Indonesia: Universitas Gajah Mada; 2018

[31] Vellend M, Geber M. Connection between species diversity and genetic diversity. Ecological Letter. 2005;**8**:767-781

[32] Jayusman J, Naiem M, Indricko S, Hardiyanto EB, Nurcahyaningsih ILG. Assessment of genetic diversity among surian *Toona sinensis* Roem in progenies test using random amplified polymorphic DNA markers. Indonesian Journal of Biotechnology. 2017;**22**(1):22-30

[33] Erikson G, Namkoong G, Roberds JH. Dynamic gene conservation for uncertain future. Forest Ecology and Management. 1993;**62**:15-37

[34] Azizah. Keragaman Genetik dan Sistem Perkawinan pada Tegakan Benih Mindi (Melia azedarach Linn.) di Wanayasa, Purwakarta [thesis]. Indonesia; (tidak dipublikasikan): Intitut Pertanian Bogor; 2014

[35] Lemes MR, Grattapaglia D, Grogan J, Proctor J. Gribel R: Flexible mating system in a logged population of Swietenia macrophylla king (Meliaceae): Implications for the management of a threatened neotropical tree species. Plant Ecology. 2007;**192**:169-179

[36] Sánchez LGH. Genetic Diversity and Mating System Analysis of Cedrela odorata L. (Meliaceae) Populations under Differ Magister Scientiae en Manejo Conservación de Bosques Naturales Biodiversidadent Human Dominated Landscapes and Primary Forests. Turrialba, Costa Rica: Tropical Agricultural Research, and Higher Education Center; 2008. 74 p

[37] Hall P, Orrell LC, Bawa KS. Genetic diversity and mating system in a tropical tree, Carapa guianensis (Meliaceae). American Journal of Botany. 1994;**81**(9):1104-1111

[38] Kundu SK. The mating system and genetic significance of polycarpy in the neem tree (Azadirachta indica). Theoretical and Applied Genetics. 1999;**99**(7-8):1216-1220

[39] Wang CJ, Cao S, Tian Y, Wang Z, Chen M, Gong G. Germplasm resourcesresearch of *Toona sinensis* with RAPD and isoenzyme analysis. Biologia. 2008;**63**:320-326

[40] White T. Breeding strategies for forest trees: Concepts and challenges. Southern African Forestry Journal. 2001;**190**:31-42

[41] Zas R, Merlo E, Lopez F. Genotype x Invironment interaction in maritime pine families in Galicia, Noethwest Spain. Silvae Genetica. 2004;**53**(4):175-182

[42] Ruotsalainen S. Increased forest production through forest tree breeding. Scandinavian Journal of Forest Research. 2014;**29**(4):333-344

[43] Jayusman. Respon Pertumbuhan Stek Surian Putih Berdasarkan Konsentrasi Hormon Pertumbuhan dan Bentuk Stek. Wana Benih. 2016;**17**(1):1-7. ISSN 1410-1173